Solutions Manual for
Introduction to
Modern Statistical Mechanics

David Wu and David Chandler

New York   Oxford
OXFORD UNIVERSITY PRESS
1988

Oxford University Press

Oxford New York Toronto
Delhi Bombay Calcutta Madras Karachi
Petaling Jaya Singapore Hong Kong Tokyo
Nairobi Dar es Salaam Cape Town
Melbourne Auckland

and associated companies in
Berlin Ibadan

Published by Oxford University Press, Inc.,
198 Madison Avenue, New York, New York 10016-4314

Library of Congress Cataloging-in-Publication Data

Wu, David.
Solutions manual for "Introduction to modern statistical mechanics".

1. Statistical mechanics—Problems, exercises, etc.
2. Statistical thermodynamics—Problems, exercises, etc.
3. Chemistry, Physical and theoretical—Problems, exercises, etc. I. Chandler, David, 1944–
II. Chandler, David, 1944– . Introduction to modern statistical mechanics. III. Title.
QC174.844.W82 1988 530.1'3 88-27824
ISBN 0-19-505889-5

9 8 7 6 5 4

Printed in the United States of America
on acid-free paper

Preface

The following pages describe the solutions to many of the Exercises in David Chandler's text, Introduction to Modern Statistical Mechanics. That text is brief but demanding with the principles illustrated only through the numerous Exercises. Especially in light of this strategy, a solution manual can serve an important pedagogical role. While we do not present herein the solutions to all the Exercises, those we have skipped require no techniques that are not already illustrated by the solutions we have included. In some cases, we have pointed out additional references. While not absolutely necessary (the text is self contained), these additions may be especially useful for the topics not covered explicitly in the main body of the text. For instance, in Chapter 6 where at least one exercise leads one naturally to perform a finite size scaling calculation, we alert the reader to some introductory literature on this important technique.

In the format we have adopted, the manual should look like neatly prepared problem sets. The prose is kept to a minimum, and the figures are generally hand drawn sketches, though a few graphs were plotted with the aid of a microcomputer. Concerning the preparation of the manuscript, we owe great thanks to Mary Hammond for her expert secretarial help.

Solution Manual for Introduction to

Modern Statistical Mechanics

Contents

1.2 Entropy is postulated by the 2nd Law to be a monotonically increasing function of E (i.e., $T \geq 0$) and an extensive function. The first equation of state satisfies both criteria, while the second is not extensive since it grows exponentially with system size. So

$$S = L_0\gamma(\theta E/L_0)^{1/2} - L_0\gamma[\tfrac{1}{2}(\tfrac{L}{L_0})^2 + \tfrac{L_0}{L} - \tfrac{3}{2}]$$

is the correct choice. Indeed, you can convince yourself that

$$S(\lambda E,\lambda L,\lambda n) = \lambda S(E,L,n) .$$

Further,

$$\frac{1}{T} = (\frac{\partial S}{\partial E})_{n,L} = \frac{1}{2}\frac{\gamma\sqrt{L_0\theta}}{\sqrt{E}}$$

implying $T > 0$. For the tension, we have

$$-f/T = (\frac{\partial S}{\partial L})_{E,n} = -L_0\gamma[\frac{L}{L_0^2} - \frac{L_0}{L^2}]$$

or $f = T\gamma(\ell/\ell_0)[1-(\ell_0/\ell)^3]$, $\ell = L/n$.

1.4 The principles we apply are conservation of total extensive energy and equality of temperature at thermal equilibrium. From the solution to Exercise 1.2,

$$E = T^2\gamma^2 n\ell_0\theta/4 .$$

Therefore, the total initial energy is

$$[(T^{(1)})^2 n^{(1)} + (T^{(2)})^2 n^{(2)}]\gamma^2\ell_0\theta/4 = E^{(1)} + E^{(2)}$$

which is the same as the total final energy

$$T^2(n^{(1)} + n^{(2)})\gamma^2\ell_0\theta/4 .$$

Hence

$$T = \{ \frac{1}{n^{(1)} + n^{(2)}} [(T^{(1)})^2 n^{(1)} + (T^{(2)})^2 n^{(2)}] \}^{1/2} .$$

1.5 Assuming the system is surrounded by adiabatic walls, the change in energy ΔE is equal to the work $\int pdV$ due to changing the pressures p_A and p_B. The final equilibrium state so obtained has new mole numbers n_1 and n_2. Instead we can imagine arriving at this final equilibrium state in two steps. First we reversibly pump the particles to their final values. Then we do work by moving the pistons to their final pressures with the walls impermeable to any particles. We know the total change in energy of the system must again be ΔE since E is a function of state. Hence, the work done by changing n_1 and n_2 can be associated with work done by changing the pressures.

1.8

$$\left(\frac{\partial C_p}{\partial p}\right)_{T,n} = T \left(\frac{\partial}{\partial p} \left(\frac{\partial S}{\partial T}\right)_{p,n}\right)_{T,n}$$

$$= T\left(\frac{\partial}{\partial T} \left(\frac{\partial S}{\partial p}\right)_{T,n}\right)_{p,n}$$

$$= T\left(\frac{\partial}{\partial T} \left(- \frac{\partial V}{\partial T}\right)_{p,n}\right)_{p,n}$$

$$= - T\left(\frac{\partial^2 V}{\partial T^2}\right)_{p,n}$$

p,T,n natural variables =>

$dG = -SdT + Vdp + \mu dn$

1.12 For a rubber band of length L, tension f,

$$dE = TdS + fdL + \mu dn .$$

E is extensive. Therefore by Euler's theorem, $E = TS + fL + \mu n$. Hence,

$$dE = TdS + SdT + fdL + Ldf + \mu dn + nd\mu$$

implying

$$0 = SdT + Ldf + nd\mu \ .$$

1.13 Since,

$$E = \theta S^2L/n^2$$

we have

$$\mu = \left(\frac{\partial E}{\partial n}\right)_{S,L} = -\frac{2\theta S^2L}{n^3} \ .$$

Substitute S = S(T,L,n) by inverting T = T(S,L,n):

$$T = \left(\frac{\partial E}{\partial S}\right)_{L,N} = \frac{2\theta SL}{n^2} \ , \quad \text{so} \quad S = \frac{n^2T}{2\theta L} \ .$$

Thus,

$$\mu = -\frac{2\theta\left(\frac{n^2T}{2\theta L}\right)^2 L}{n^3} = -\frac{T^2}{2\theta(L/n)} \ .$$

The Gibbs-Duhem like equation is $0 = SdT + Ldf + nd\mu$. We want to show this interdependence of these three differentials. We choose a representation, e.g., $\{S,L,n\}$, though we could also use $\{T,f\}$:

$$dT = \left(\frac{\partial T}{\partial S}\right)_{L,n} dS + \left(\frac{\partial T}{\partial L}\right)_{S,n} dL + \left(\frac{\partial T}{\partial n}\right)_{S,L} dn$$

$$= \left(\frac{2\theta L}{n^2}\right)dS + \left(\frac{2\theta S}{n^2}\right)dL + \left(-\frac{4\theta SL}{n^3}\right)dn \ .$$

Similarly for df and dμ

$$f = \left(\frac{\partial E}{\partial L}\right)_{S,n} = \theta\frac{S^2}{n^2}$$

$$df = \left(\frac{2\theta S}{n^2}\right)dS + \left(-\frac{2\theta S^2}{n^3}\right)dn$$

$$d\mu = \left(-\frac{4\theta SL}{n^3}\right)dS + \left(-\frac{2\theta S^2}{n^3}\right)dL + \left(\frac{6\theta S^2L}{n^4}\right)dn \ .$$

Plugging into $0 = SdT + Ldf + nd\mu$

$$[S(\frac{2\theta L}{n^2}) + L(\frac{2\theta S}{n^2}) + n(-\frac{4\theta SL}{n^3})]\ dS$$

$$+ [S(\frac{2\theta S}{n^2}) + L(0) + n(-\frac{2\theta S^2}{n^3})]\ dL$$

$$+ [S(-\frac{4\theta SL}{n^3} + L(-\frac{2\theta S^2}{n^3}) + n(\frac{6\theta S^2 L}{n^4})\ dn$$

$$= 0 \ .$$

QED

1.14 The Gibbs-Duhem equation, $0 = -SdT + Vdp - nd\mu$, implies

$$d\mu = -sdT + vdp \ , \quad \text{with } s = S/n \text{ and } v = V/n \ .$$

Hence,

$$(\frac{\partial\mu}{\partial v})_T = -s(\frac{\partial T}{\partial v})_T + v(\frac{\partial p}{\partial v})_T$$

$$= v(\frac{\partial p}{\partial v})_T \ .$$

1.16 From the definition of heat capacity

$$(\frac{\partial c_\ell}{\partial \ell})_T = (\frac{\partial}{\partial \ell}\, T(\frac{\partial s}{\partial T})_\ell)_T = T(\frac{\partial}{\partial T}(\frac{\partial s}{\partial \ell})_T)_\ell \ .$$

Further, since $dE = TdS + fdl + \mu dn$, we have

$$d(E-TS) = -\,SdT + fdL + \mu dn$$

and thus the Maxwell-type relation

$$(\frac{\partial s}{\partial \ell})_T = -\,(\frac{\partial f}{\partial T})_\ell$$

Hence

$$(\frac{\partial c_\ell}{\partial \ell})_T = -\,T(\frac{\partial^2 f}{\partial T^2})_\ell = -\,T\,\frac{\partial^2}{\partial T^2}\,(\frac{\ell T}{\theta}) = 0 \ .$$

2.9 (a) Stability criteria are generally $(\frac{\partial I_i}{\partial X_i}) \geq 0$ where X_i and I_i are conjugate. Therefore (i) is true,

$$(\frac{\partial p}{\partial v})_T = n(\frac{\partial p}{\partial V})_{T,n} < 0 \ , \quad \text{since } -p \text{ and } V \text{ are conjugate.}$$

Also (iii) is true since $(\frac{\partial \mu}{\partial v})_T = + v(\frac{\partial p}{\partial v})_T < 0$ by the Gibbs-Duhem equation. None of the remainder are guaranteed false by stability.

(b) From Part (a), we know to examine (ii) and (iv). Note

$$(\frac{\partial T}{\partial v})_s = - (\frac{\partial T}{\partial s})_v (\frac{\partial s}{\partial v})_T$$

$$= - (\frac{\partial T}{\partial s})_v (\frac{\partial p}{\partial T})_v \ . \qquad \{dA = -SdT - pdV + \mu dN\}$$

Since $(\partial T/\partial s)_v$ is positive by stability,

$$(\frac{\partial T}{\partial v})_s > 0 \text{ implies } (\frac{\partial p}{\partial T})_v < 0 \ .$$

But this implication contradicts (ii). Thus, (ii) and (iv) are inconsistent.

2.10 To derive the analog of the Clausius-Clapeyron equation in the μ-T plane, note that the μ-T coexistence line requires that $p^{(\alpha)}(\mu,T) = p^{(\beta)}(\mu,T)$. But $0 = -SdT + Vdp - nd\mu$ implies $dp + (S/V)dT + (n/V)d\mu$. As we creep infinitesimally along the μ-T coexistence line, $dp^{(\alpha)} = dp^{(\beta)}$ when infinitesimally close on either side, i.e.,

$$(\frac{S}{V})^{(\alpha)}dT + (\frac{n}{V})^{(\alpha)}d\mu = (\frac{S}{V})^{(\beta)}dT + (\frac{n}{V})^{(\beta)}d\mu$$

or

$$\frac{dT}{d\mu} = - \frac{\Delta\rho(T)}{\Delta\hat{s}(T)} \ ,$$

where

$$\Delta\rho = \left(\frac{n}{V}\right)^{(\alpha)} - \left(\frac{n}{V}\right)^{(\beta)} ,$$

$$\Delta\hat{s} = \left(\frac{S}{V}\right)^{(\alpha)} - \left(\frac{S}{V}\right)^{(\beta)} .$$

2.19 $C_v = T\left(\frac{\partial S}{\partial T}\right)_v \qquad C_p = T\left(\frac{\partial S}{\partial T}\right)_p \qquad K_s = -\frac{1}{V}\left(\frac{\partial V}{\partial p}\right)_s \qquad K_T = -\frac{1}{V}\left(\frac{\partial V}{\partial p}\right)_T$

Assume constant n . To prove $(K_s/K_T) = (C_v/C_p)$ we note that

$$\begin{aligned} K_s/K_T &= \left(\frac{\partial V}{\partial p}\right)_s / \left(\frac{\partial V}{\partial p}\right)_T \\ &= \left(\frac{\partial V}{\partial S}\right)_p \left(\frac{\partial S}{\partial p}\right)_v \left(\frac{\partial p}{\partial T}\right)_v \left(\frac{\partial T}{\partial V}\right)_p \\ &= \left(\frac{\partial S}{\partial T}\right)_v \left(\frac{\partial T}{\partial S}\right)_p = \left(\frac{\partial S}{\partial T}\right)_v / \left(\frac{\partial S}{\partial T}\right)_p = \frac{C_v}{C_p} . \end{aligned}$$

The second equality comes from realizing $\left(\frac{\partial x}{\partial y}\right)_z = -\left(\frac{\partial x}{\partial z}\right)_y \left(\frac{\partial z}{\partial y}\right)_x$, and the third uses the chain rule. Since

$$C_p - C_v = -T\left(\frac{\partial p}{\partial V}\right)_{T,n} \left[\left(\frac{\partial V}{\partial T}\right)_{p,n}\right]^2$$

and since $-(\partial p/\partial v)_T > 0$ by stability, we have $C_p > C_v$. Thus, $C_p/C_v > 1$, and since $C_p/C_v = K_T/K_s$, we have $K_T > K_s$. This inequality says that instead of working adiabatically, it is easier to compress something if you can leak energy out of the system into a constant temperature heat bath.

2.20 (a) The rubber band heats up when stretched. Assume constant n. So,

$$\left(\frac{\partial T}{\partial L}\right)_{S,n} > 0 \quad \text{or equivalently} \quad \left(\frac{\partial T}{\partial f}\right)_{S,n} > 0 .$$

Note, these two derivatives have the same sign since

$$\left(\frac{\partial T}{\partial f}\right)_{S,n} = \left(\frac{\partial T}{\partial L}\right)_{S,n} \left(\frac{\partial L}{\partial f}\right)_{S,n}$$

and the second derivative is positive by stability.

To find the sign of $\left(\frac{\partial L}{\partial T}\right)_{f,n}$, we write

$$\left(\frac{\partial L}{\partial T}\right)_{f,n} = \left(\frac{\partial L}{\partial S}\right)_{f,n} \left(\frac{\partial S}{\partial T}\right)_{f,n}$$

$$= -\underbrace{\left(\frac{\partial T}{\partial f}\right)_S}_{\substack{>0 \\ \text{given}}} \underbrace{\left(\frac{\partial S}{\partial T}\right)_f}_{\substack{>0 \\ \text{by stability}}}$$

$d(E-fL) = TdS - Ldf + \ldots$ implies $\left(\frac{\partial L}{\partial S}\right)_{f,n} = -\left(\frac{\partial T}{\partial f}\right)_{S,n}$

Thus

$$\left(\frac{\partial L}{\partial T}\right)_{f,n} < 0 .$$

In other words, the rubber band stretches when cooled.

(b) $$\left(\frac{\partial S}{\partial T}\right)_f = \left(\frac{\partial S}{\partial T}\right)_L \left(\frac{\partial T}{\partial T}\right)_f + \left(\frac{\partial S}{\partial L}\right)_T \left(\frac{\partial L}{\partial T}\right)_f$$

$$\left(\frac{\partial S}{\partial T}\right)_f - \left(\frac{\partial S}{\partial T}\right)_L = \left(\frac{\partial S}{\partial L}\right)_T \left[-\left(\frac{\partial f}{\partial T}\right)_L \left(\frac{\partial L}{\partial f}\right)_T\right]$$

$$= \underbrace{\left(\frac{\partial s}{\partial L}\right)_T^2}_{>0} \underbrace{\left(\frac{\partial L}{\partial f}\right)_T}_{\substack{>0 \\ \text{by stability}}}$$

$d(E - TS) = -SdT + fdL + \ldots$ implies $\left(\frac{\partial f}{\partial T}\right)_L = -\left(\frac{\partial S}{\partial L}\right)_T$

$$C_f - C_L > 0 .$$

Hence $\frac{1}{C_L} > \frac{1}{C_f}$. In other words, the constant length rubber band has the larger change in temperature.

2.21(a) Since in general $dE = TdS + \underset{\sim}{f}\cdot d\underset{\sim}{x}$, where $\underset{\sim}{f}$ is the intensive field and $\underset{\sim}{x}$ is the extensive parameter, the correct work term is HdM. (In other words, doubling the system doubles M, not H. This approximate extensivity neglects interactions such as mutual polarization which might occur if two magnetizable systems were placed adjacent to each other.) So

$$dE = TdS + HdM - pdV + \mu dN ,$$

and

$$dA = -SdT + HdM - pdV - \mu dN .$$

Stability then requires both $\chi_S^{-1} = \left(\frac{\partial H}{\partial M}\right)_{S,V,n} > 0$ and

$$\chi_T^{-1} = \left(\frac{\partial H}{\partial M}\right)_{T,V,n} > 0 \; .$$

Using the representation $\{S,H,n,V\}$, the thermodynamic potential is $E - HM$. We then have $M = M(S,H,n,V)$ since $M = \left(\frac{\partial(E-HM)}{\partial H}\right)_{S,n,V}$.

$$\chi_T = \left(\frac{\partial M}{\partial H}\right)_T = \left(\frac{\partial M}{\partial S}\right)_H \cdot \left(\frac{\partial S}{\partial H}\right)_T + \left(\frac{\partial M}{\partial H}\right)_S$$

$$\chi_T - \chi_S = \left(- \frac{\partial T}{\partial H}\right)_S \cdot \left(-\frac{\partial T}{\partial H}\right)_S \cdot \left(\frac{\partial S}{\partial T}\right)_H$$

$$= \left(\frac{\partial T}{\partial H}\right)_S^2 / \left(\frac{\partial T}{\partial S}\right)_H > 0$$

$d(E-HM) = TdS - MdH + \ldots$

By the Maxwell relation

$$\left(\frac{\partial M}{\partial S}\right)_H = -\left(\frac{\partial T}{\partial H}\right)_S \; .$$

$\left(\frac{\partial T}{\partial S}\right)_H > 0$ by stability.

(b) Given $\left(\frac{\partial M}{\partial T}\right)_H < 0$, we want to determine the sign of $\left(\frac{\partial T}{\partial H}\right)_S$:

$$\left(\frac{\partial T}{\partial H}\right)_S = - \left(\frac{\partial T}{\partial S}\right)_H \cdot \left(\frac{\partial S}{\partial H}\right)_T$$

$$= - \underbrace{\left(\frac{\partial T}{\partial S}\right)_H}_{>0,\text{ by stability}} \underbrace{\left(\frac{\partial M}{\partial T}\right)_H}_{<0,\text{ given}}$$

$d(E-HM-ST) = -SdT - MdH + \ldots$

By the Maxwell relation

$$\left(\frac{\partial S}{\partial H}\right)_T = \left(\frac{\partial M}{\partial T}\right)_H \; .$$

2.23 (a) $dA = -SdT + fdL + \mu dM$ $\qquad x = L/M$

k, h, x_0 and c are independent of x, but not T.

$$f_S = \left(\frac{\partial A_S}{\partial L}\right)_{T,M} = Mkx \cdot \frac{1}{M} = kx$$

$$f_L = Mh(x-x_0) \cdot \frac{1}{M} = h(x-x_0)$$

(b) $$A_S = \frac{1}{2} kMx^2 = \frac{1}{2} kL^2/M$$

$$\mu_S = \left(\frac{\partial A_S}{\partial M}\right)_{T,L} = - \frac{kL^2}{2M^2} = - \frac{1}{2} kx^2$$

$$A_L = \frac{1}{2} hM\left(\frac{L}{M} - x_0\right)^2 + cM = \frac{1}{2} h\left[\frac{L^2}{M} - 2Lx_0 + Mx_0^2\right] + cM$$

Since x_0 is independent of L and of x, it's independent of M.

$$\mu_L = \frac{1}{2} h\left[- \frac{L^2}{M^2} + x_0^2\right] + c = - \frac{1}{2} h(x^2 - x_0^2) + c$$

(c) $$\frac{A_S}{M} - f_S x = - \frac{1}{2} kx^2 = \mu_S$$

$$\frac{A_L}{M} - f_L x = -\frac{1}{2} h(x^2 - x_0^2) + c = \mu_L$$

(d) The phase transition occurs at $\mu_S = \mu_L = \mu_T$, $f_S = f_L = f_T$. At μ_T, f_T , the short phase has $x = x_S$, and the long phase has $x = x_L$.

$\mu_S = \mu_L$ implies $-\frac{k}{2} x_S^2 = -\frac{h}{2} (x_L^2 - x_0^2) + c$.

$f_S = f_L$ implies $kx_S = h(x_L - x_0)$.

Thus

$$(kx_S)^2 = hkx_L^2 - hkx_0^2 - 2ck ,$$

or

$$x_L - x_0 = \frac{hkx_0 \pm \sqrt{hk}\,\sqrt{hkx_0^2 + 2c(k-h)}}{h(h-k)} .$$

Since

$$f_T = h(x_L - x_0)$$

and

since $k>h$, $c>0$, we have,

$$f_T = \frac{\sqrt{hk}\,\sqrt{hkx_0^2 + 2c(k-h)} - hkx_0}{k-h} .$$

(e) $x_S = \frac{h}{k} (x_L - x_0)$ implies $\Delta x = x_L - x_S = (\frac{k-h}{k})(x_L - x_0) + x_0$

or

$$\Delta x = \sqrt{h/k}\;\sqrt{hkx_0^2 + 2c(k-h)} - hx_0 + x_0 .$$

2.24 Near the liquid-solid phase coexistence

$$A^{(\ell)} = \frac{1}{2}\frac{\alpha}{T}\frac{n^2}{V} , \qquad A^{(s)} = \frac{1}{3}\frac{\beta}{T}\frac{n^3}{V^2}$$

which are functions in $\{n,V,T\}$.

(a) $\rho^{(\ell)}$ and $\rho^{(s)}$ are the densities near the coexistence line. At coexistence, the temperature is the remaining free parameter,

$$\mu^{(s)} = \mu^{(\ell)} \quad , \quad p^{(s)} = p^{(\ell)} \quad , \quad T^{(s)} = T^{(\ell)} = T \ .$$

Explicitly,

$$\left(\frac{\partial A^{(s)}}{\partial n}\right)_{V,T} = \left(\frac{\partial A^{(\ell)}}{\partial n}\right)_{V,T} \quad , \quad -\left(\frac{\partial A^{(s)}}{\partial V}\right)_{n,T} = -\left(\frac{\partial A^{(\ell)}}{\partial V}\right)_{n,T}$$

So

$$\frac{\beta}{T}\left[\left(\frac{n}{V}\right)^{(s)}\right]^2 = \frac{\alpha}{T}\left(\frac{n}{V}\right)^{(\ell)} \quad , \quad \frac{2\beta}{3T}\left[\left(\frac{n}{V}\right)^{(s)}\right]^3 = \frac{\alpha}{2T}\left[\left(\frac{n}{V}\right)^{(\ell)}\right]^2 \ .$$

Substituting $\frac{\beta}{\alpha}[\rho^{(s)}]^2 = \rho^{(\ell)}$ into the previous equation we get

$$[\rho^{(s)}]^3 = \frac{3\alpha}{4\beta}\left[\frac{\beta}{\alpha}\,\rho^{(s)2}\right]^2 \ .$$

Therefore the densities are

$$\rho^{(s)} = \frac{4}{3}\,\frac{\alpha}{\beta} \quad \text{and} \quad \rho^{(\ell)} = \frac{16}{9}\,\frac{\alpha}{\beta} \ ,$$

and are constant functions of T.

(b) $$p_s = p^{(s)} = p^{(\ell)} = \frac{\alpha}{2T}\left(\rho^{(\ell)}\right)^2 = \frac{128}{81}\,\frac{\alpha^3}{\beta^2}\,\frac{1}{T}$$

(c) $$S^{(\ell)} = -\left(\frac{\partial A^{(\ell)}}{\partial T}\right)_{n,V} = +\frac{1}{2}\,\frac{\alpha}{T^2}\,\frac{n^2}{V} \ , \quad S^{(s)} = -\left(\frac{\partial A^{(s)}}{\partial T}\right)_{n,V} = +\frac{1}{3}\,\frac{\beta}{T^2}\,\frac{n^3}{V^2}$$

$$\Delta s_{\ell\to s} = \frac{S^{(s)} - S^{(\ell)}}{n} = \frac{1}{T^2}\left(\frac{\beta}{3}\,\rho^{(s)2} - \frac{\alpha}{2}\,\rho^{(\ell)}\right) = \frac{1}{T^2}\left[\frac{16}{27}\,\frac{\alpha^2}{\beta} - \frac{8}{9}\,\frac{\alpha^2}{\beta}\right]$$

$$= -\frac{8}{27}\,\frac{1}{T^2}\,\frac{\alpha^2}{\beta} \qquad \text{where } s = S/n$$

(d) Since $\Delta v = \frac{3}{16}\,\frac{\beta}{\alpha}$,

$$\frac{\Delta s}{\Delta v} = -\frac{128}{81}\,\frac{\alpha^3}{\beta^2}\,\frac{1}{T^2} \ .$$

Also,

$$\frac{dp_s}{dT} = -\frac{128}{81}\,\frac{\alpha^3}{\beta^2}\,\frac{1}{T^2} \ .$$

Therefore $\frac{\Delta s}{\Delta v} = \frac{dp}{dT}$, satisfying the Clausius-Clapeyron equation.

2.25 Stability requires $-\left(\frac{\partial p}{\partial V}\right)_{T,n} > 0$

or $\left(\frac{\partial p}{\partial \rho}\right)_{T,n} = -\frac{V^2}{n}\left(\frac{\partial p}{\partial V}\right)_{T,n} > 0$.

From the van der Waals equation,

$$\left(\frac{\partial p}{\partial \rho}\right)_{T,n} = \frac{RT}{(1-b\rho)^2} - 2a\rho \quad .$$

Instability occurs when this function is negative. The boundary surrounding the region of instability, the spinodal, is where it's 0: i.e.,

$$RT - 2a\rho(1-b\rho)^2 = 0 \ .$$

In the T-ρ plane, the spinodal is $RT = 2a\rho(1-b\rho)^2$, i.e., the spinodal is a cubic polynomial with roots for $T = 0$ at $\rho = 0,\ 1/b$. Also note that the spinodal has extrema when

$$\frac{\partial T}{\partial \rho} = 0 = 3b^2\rho^2 - 4b\rho + 1 \quad ,$$

that is when $\rho = 1/b\ ,\ \ 1/3b$.

Similarly, there is a point of inflection when

$$\frac{d^2T}{\partial \rho^2} = 0 = 3b^3\rho - 2b = 0 \quad ,$$

that is when $\rho = 2/3b$.

Finally,

$$T(\rho = \frac{1}{3b}) = \frac{8}{27}\frac{a}{b} \ .$$

These features are illustrated in the figure below. Only $\rho < b^{-1}$ is considered since higher ρ would give negative pressures. Note also that the pressure on the spinodal

$$p_{spinodal} = p[T_{spinodal}(\rho),\rho] = a\rho^2(1-2b\rho)$$

becomes negative for $\rho > 1/2b$.

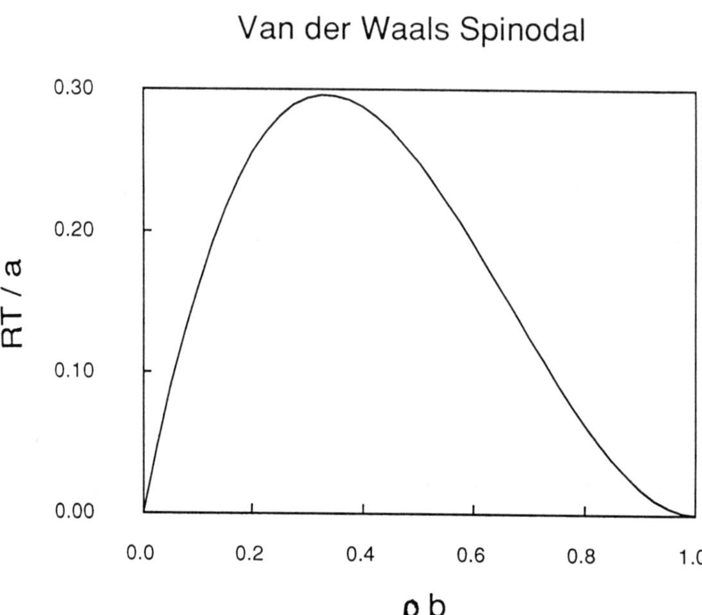

The van der Waals equation of state for a given temperature T_0 has only one interval in ρ, if at all, for which $(\frac{\partial p}{\partial \rho})_T < 0$, i.e., for which the fluid is unstable. This is reflected by the fact that the spinodal equation, $p = a\rho^2(1-2b\rho)$, has only two distinct roots.

In other words, the isotherm is shaped as illustrated in the figure on the next page. The Maxwell construction requires $p(\rho_g) = p(\rho_\ell)$ which can only be satisfied within the height h. Any pair of points so picked will lead to a coexistence curve enveloping the spinodal region.

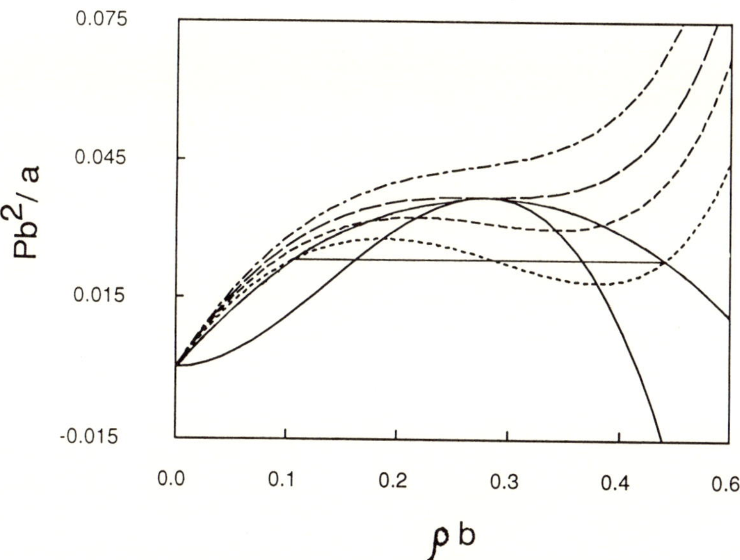

2.26 phase α: $\beta p = a + b\beta\mu$

phase β: $\beta p = c + d(\beta\mu)^2$

At phase equilibrium,

$$\beta p^{(\alpha)} = \beta p^{(\beta)} \;, \quad \beta\mu^{(\alpha)} = \beta\mu^{(\beta)} \;, \quad \text{and } \beta_\alpha^{(\alpha)} = \beta^{(\beta)} = \beta \;.$$

Thus $a + b\beta\mu = c + d(\beta\mu)^2$, implying

$$\beta\mu = \frac{b \pm \sqrt{b^2 - 4d(c-a)}}{2d} \;.$$

The density can be obtained from the Gibbs-Duhem equation,

$$d\mu = -sdT + vdp \;, \text{ i.e., } (1/v) = \rho = \left(\frac{\partial p}{\partial \mu}\right)_T = \left(\frac{\partial(\beta p)}{\partial(\beta\mu)}\right)_\beta \;.$$

So, $\rho^{(\alpha)} = b$,

$\rho^{(\beta)} = 2d\beta\mu$,

and we identify the positive root as the physical root to the quadratic equation. Hence

$$\rho^{(\beta)} - \rho^{(\alpha)} = \sqrt{b^2 + 4d(a-c)}$$

and

$$\beta p_{transition} = a + \frac{b}{2d} [b + \sqrt{b^2 + 4d(a-c)}] .$$

3.8 The probability in a canonical ensemble is

$$P(E) \propto \Omega(E)e^{-\beta E} = \exp[\ln\Omega(E)-\beta E] \ .$$

We assume that P(E) is so sharply peaked that $\langle E\rangle$ occurs at the top (or very close to the top) of the peak. The following steepest descent calculation will validate our assumption: Expand $\ln P(E) = \ln\Omega(E) - \beta E$ about its maximum value $\ln P(\langle E\rangle)$ which occurs at $\langle E\rangle$, i.e.,

$$\ln P(E) = \ln P(\langle E\rangle) + \delta E \underbrace{\frac{\partial \ln P(E)}{\partial E}\Big|_{\langle E\rangle}}_{\substack{0,\text{ because } \ln P(E) \\ \text{has a maximum at } \langle E\rangle}} + \frac{1}{2}(\delta E)^2 \underbrace{\frac{\partial^2 \ln P(E)}{\partial E^2}\Big|_{\langle E\rangle}}_{\text{negative}} + \cdots$$

$$\frac{\partial^2 \ln P}{\partial E^2}\Big|_{\langle E\rangle} = \frac{\partial^2(\ln\Omega-\beta E)}{\partial E^2}\Big|_{\langle E\rangle} = \frac{\partial^2 \ln\Omega}{\partial E^2}\Big|_{\langle E\rangle}$$

$$= \left(\frac{\partial}{\partial E}\left(\frac{\partial \ln\Omega}{\partial E}\right)\right)_{\langle E\rangle} = \frac{\partial}{\partial\langle E\rangle}\left(\frac{\partial \ln\Omega(\langle E\rangle)}{\partial\langle E\rangle}\right) = \frac{\partial\beta}{\partial\langle E\rangle}$$

$$= -\frac{1}{k_B T^2}\frac{\partial T}{\partial\langle E\rangle} = -\frac{1}{k_B T^2 C_V} \ .$$

Hence

$$P(E) \propto \exp\left[-\frac{(E-\langle E\rangle)^2}{2k_B T^2 C_V}\right]$$

For $\delta E \approx 10^{-6}\langle E\rangle$, $N = 10^{21}$, using $\langle E\rangle = 3/2\, Nk_B T$, $C_V = 3/2\, Nk_B$,

$$P(E)/P(\langle E\rangle) \approx e^{-10^9} \approx \underbrace{0.000.....0001}_{\text{a billion zeros}} \ .$$

That's improbable!

3.9 $$m/N = \frac{1}{1+e^{\beta\varepsilon}} \text{ implies } \frac{N}{N-m} = \frac{1}{1-m/N} = \frac{1+e^{\beta\varepsilon}}{e^{\beta\varepsilon}} = e^{-\beta\varepsilon}+1 \ ,$$

and

$$\frac{N-m}{m} = N/m - 1 = e^{\beta\varepsilon} \ .$$

So

$$S/k_B = \ln\Omega(E,N) = \ln\left(\frac{N!}{(N-m)!m!}\right) \qquad \text{where } m = m(E) = E/\varepsilon$$

$$= N \ln N - (N-m) \ln (N-m) - m \ln m$$

$$= N \ln \left(\frac{N}{N-m}\right) + m \ln \left(\frac{N-m}{m}\right)$$

$$= N \ln (e^{-\beta\varepsilon} + 1) + \frac{N}{1 + e^{\beta\varepsilon}} \beta\varepsilon = S(\beta,N)/k_B$$

Clearly, $\lim_{\beta\to\infty} S/k_B = 0$.

Also,

$$S/k_B(E,N) = N \ln \left(\frac{N}{N-m(E)}\right) + m(E) \ln \left(\frac{N}{m(E)} - 1\right)$$

$$= - N \ln \left(1 - \frac{E}{N\varepsilon}\right) + E/\varepsilon \ln \left(\frac{N\varepsilon}{E} - 1\right) .$$

So $\quad 1/k_BT = \left.\frac{\partial(S/k_B)}{\partial E}\right|_N = \frac{1}{\varepsilon} \ln \left(\frac{N\varepsilon}{E} - 1\right)$.

The last result is in agreement with the earlier formula, $N/m - 1 = e^{\beta\varepsilon}$. Notice, however, that $1/T$ can be negative when $\frac{N\varepsilon}{E} - 1 < 1$, i.e., $\frac{E}{N} > \frac{\varepsilon}{2}$ implies $\frac{1}{T} < 0$.

3.10 From p. 68, the canonical ensemble calculation gives

$$-\beta A = N \ln(1 + e^{-\beta\varepsilon}) \qquad \langle E\rangle = N\varepsilon(1 + e^{\beta\varepsilon})^{-1}$$

Then

$$S/k_B = -\beta(A - \langle E\rangle) = N \ln(1 + e^{-\beta e}) + \frac{N\beta\varepsilon}{1 + e^{\beta\varepsilon}} ,$$

which is the same as in Exercise 3.9 from the microcanonical calculation.

3.16 $N = 0.01$ moles is $\approx 10^{22}$. An open, thermally equilibrated system is controlled by T, V and μ. Fluctuations in the energy and the density $\rho = N/v$ are given by

$$\langle(\delta x)^2\rangle = - \frac{\partial\langle x\rangle}{\partial\xi} , \qquad \text{where } \xi \text{ is conjugate to } x .$$

For E and N varying, the grand canonical ensemble yields

$$\langle(\delta E)^2\rangle = -\left(\frac{\partial\langle E\rangle}{\partial\beta}\right)_{V,\beta\mu} = k_B T^2 \left(\frac{\partial\langle E\rangle}{\partial T}\right)_{V,\beta\mu} .$$

We could calculate this last derivative precisely (see below), but for the purposes of an estimate, we only need to know that we'd only get some constant times C_V, the exact constant being irrelevant since it's still O(N). So, as an estimate

$$\langle(\delta E)^2\rangle \approx k_B T^2 C_V$$

which is the canonical ensemble result. The RMS deviation

$$\frac{\sqrt{\langle(\delta E)^2\rangle}}{\langle E\rangle} = \frac{\sqrt{k_B T^2 3k_B N/2}}{3k_B TN/2} \approx \frac{1}{\sqrt{N}} \approx 10^{-11} .$$

Similarly, for an ideal gas, $\langle(\delta N)^2\rangle = \langle N\rangle$

$$\text{So } \frac{\sqrt{\langle(\delta p)^2\rangle}}{\langle p\rangle} = \frac{\sqrt{\langle(\delta N)^2\rangle}}{\langle N\rangle} = \frac{1}{\sqrt{N}} = 10^{-11} .$$

Just for reference, let's evaluate the actual derivatives involved in $\langle(\delta E)^2\rangle$:

$$\left(\frac{\partial E}{\partial T}\right)_{V,\beta\mu} = \left(\frac{\partial E}{\partial T}\right)_{V,n} + \left(\frac{\partial E}{\partial N}\right)_{V,T}\left(\frac{\partial N}{\partial T}\right)_{V,\beta\mu}$$

$$= C_v + (3/2\ k_B T)\cdot\left(-\frac{\partial\beta\mu}{\partial T}\right)_{V,N} / \left(\frac{\partial\beta\mu}{\partial N}\right)_{V,T} .$$

$$\left(\frac{\partial\beta\mu}{\partial T}\right)_{V,N} = \frac{E}{N}\left(\frac{\partial\beta}{\partial T}\right)_{V,N} + \frac{V}{N}$$

$$= \frac{E}{N}\cdot\left(-\frac{1}{k_B T^2}\right) = -\frac{3}{2T}$$

$$S = 1/T\ dE + p/T\ dV - \mu/T\ dN$$

$$0 = E d\beta + V d(\beta p) - N d(\beta\mu)$$

$$\Rightarrow d(\beta\mu) = E/N\ d\beta + V/N\ d(\beta p)$$

and

$$\left(\frac{\partial\beta\mu}{\partial N}\right)_{V,T} = \frac{E}{N} + \frac{V}{N}\frac{\partial(\beta p)}{\partial N}\Big|_{V,T}$$

$$= \frac{V}{N}\cdot\frac{1}{V} = \frac{1}{N} .$$

$$\beta p = \frac{N}{V} = \beta p(N,V,T)$$

So,

$$\left(\frac{\partial E}{\partial T}\right)_{V,\beta\mu} = C_v + (3/2\ k_B T)\cdot\left(\frac{3}{2T}\right)\cdot N = (3/2 + 9/4)Nk_B = 15/4\ Nk_B .$$

3.17 $a \le x \le b$. Show $[\forall g,f | \langle gf\rangle = \langle g\rangle\langle f\rangle] \Leftrightarrow p(x) = \delta(x-x_0)$.

(a)

$$\langle gf\rangle - \langle g\rangle\langle f\rangle = \int_a^b g(x)f(x)p(x)dx - \int_a^b g(x)p(x)dx \cdot \int_a^b f(y)p(y)dy$$

$$= \int_a^b\int_a^b g(x)f(y)p(x)\delta(y-x)dxdy - \int_a^b\int_a^b g(x)f(y)p(x)p(y)dxdy \ .$$

Hence, if $\langle gf\rangle = \langle g\rangle\langle f\rangle$,

$$\int_a^b\int_a^b g(x)f(y)p(x)[\delta(y-x) - p(y)]\, dxdy = 0 \ .$$

This holds for all g and f if and only if

$$p(x)[\delta(y-x) - p(y)] = 0 \ .$$

The trivial solution, $p(x) = 0$, is discounted because it is not normalized. Thus, we must have

$$p(y) = \delta(y-x_0) \ ,$$

where x_0 is some point in the interval $[a,b]$.

(b) $\langle f(x)g(y)\rangle - \langle f(x)\rangle\langle g(y)\rangle$

$$= \int\int dxdy\ f(x)g(y)p(x,y) - \int dx\ f(x)\overline{p}_1(x) \int dy\ g(y)\overline{p}_2(y) \ .$$

Hence, if $\langle fg\rangle = \langle f\rangle\langle g\rangle$,

$$\int\int dxdy\ f(x)g(y)[p(x,y) - \overline{p}_1(x)\ \overline{p}_2(y)] = 0 \ .$$

This holds for all f and g if and only if

$$p(x,y) = \overline{p}_1(x)\ \overline{p}_2(y) \ .$$

Note: The above proofs rely upon the fact that $\int f(x)\eta(x)dx = 0$ for all f is equivalent to $\eta(x) = 0$ on the interval of integration. This is true because if $\eta(x) \ne 0$ we could pick $f(x) = \eta(x)$ for which $\int f(x)\eta(x)dx > 0$. For two dimensions, we can iterate the above theorem on

$$\int dx\, f(x)[\int dy\, g(y)\, \eta(x,y)] = 0$$

calling the term in the square brackets $\eta(x)$.

3.18 For N spins in a field H,

$$E_{n_1,n_2,\ldots n_N} = \sum_{i=1}^{N} (-n_i \mu H) \quad , \qquad n_i = \pm 1$$

(a) β, H, and N are constant in our ensemble:

$$Q = \sum_{n_1,n_2,\ldots n_N} e^{-\beta E_{n_1,n_2,\ldots n_N}} = \sum_{n_1,n_2,\ldots n_N} e^{-\beta \sum_{i=1}^{N} (-n_i \mu H)}$$

$$= \sum_{n_1,n_2,\ldots n_N} \prod_{i=1}^{N} e^{\beta n_i \mu H} = \prod_{i=1}^{N} \sum_{n=\pm 1} e^{\beta \mu H n} ,$$

since the factor for each spin is identical, so

$$Q = (e^{\beta\mu H} + e^{-\beta\mu H})^N = \left(2 \cosh(\beta\mu H)\right)^N .$$

$$\langle E\rangle = \frac{\partial \ell n Q}{\partial(-\beta)}$$

$$= N\mu H \frac{e^{-\beta\mu H} - e^{\beta\mu H}}{e^{-\beta\mu H} + e^{\beta\mu H}} = N\mu H \tanh(-\beta\mu H)$$

$$= - N\mu H \tanh(\beta\mu H)$$

(b) $$\ell n Q = \frac{S}{k_B} - \frac{E}{k_B T} = - \beta A$$

$$S = \frac{-A + E}{T}$$

$$= k_B \ell n Q + k_B \beta \langle E\rangle$$

$$= N k_B [\ell n(e^{\beta\mu H} + e^{-\beta\mu H}) - \beta\mu H \tanh(\beta\mu H)]$$

(c) In the limit $T \to 0$ or $\beta \to \infty$, $\tanh(\beta\mu H) \to 1$. So

$$\langle E\rangle_{T\to 0} = - N\mu H$$

i.e., the ground state has all spins + , or aligned with the field

$$\lim_{\beta\to\infty} Nk_B[\ln(e^{\beta\mu H} + e^{-\beta\mu H}) - \beta\mu H \tanh(\beta\mu H)] = Nk_B[\beta\mu H - \beta\mu H] = 0 ,$$

hence,

$$S_{T\to 0} = 0 .$$

3.19 (a) $\langle M\rangle = \langle \sum_{i=1}^{N} \mu n_i\rangle = -\frac{1}{H}\langle \sum_{i=1}^{N} - \mu n_i H\rangle = -\frac{\langle E\rangle}{H}$

$$= N\mu \tanh(\beta\mu H)$$

(b) $\langle(\delta M)^2\rangle = \langle(M-\langle M\rangle)^2\rangle = \langle M^2\rangle - \langle M\rangle^2 = \frac{1}{Q}\frac{\partial^2 Q}{\partial(\beta H)^2} - \frac{1}{Q^2}\left(\frac{\partial Q}{\partial(\beta H)}\right)^2$

$$\frac{1}{Q}\frac{\partial^2 Q}{\partial(\beta H)^2} = \frac{1}{Q}\frac{\partial}{\partial(\beta H)}[N\mu\, Q \tanh(\beta\mu H)]$$

$$= N\mu^2\{(N-1)\tanh^2(\beta\mu H) + 1\}$$

Hence,

$$\langle(\delta M)^2\rangle = N\mu^2\{1 - \tanh^2(\beta\mu H)\} = N\mu^2 \operatorname{sech}^2(\beta\mu H) .$$

Also,

$$\left.\frac{\partial\langle M\rangle}{\partial H}\right|_{\beta,N} = \beta N\mu^2\{1 - \tanh^2(\beta\mu H)\} ,$$

so $\left(\frac{\partial\langle M\rangle}{\partial(\beta H)}\right)_{\beta,N} = \left(\frac{\partial^2 \ln Q}{\partial(\beta H)^2}\right)_{\beta,N} = \langle(\delta M)^2\rangle .$

(c) As $\beta\to\infty$, $\tanh(\beta\mu H)\to 1$.

Therefore $\langle M\rangle_{T\to\infty} = N\mu$, and $\langle(\delta M)^2\rangle_{T\to\infty} = N\mu^2(1-1) = 0$.
In other words, the ground state with all the spins aligned has no fluctuations.

3.20 In an ensemble with constant M, all members have the same number of spins up. For a given H, this ensemble is the microcanonical ensemble since $E = -HM$ is also fixed. Note we can't generalize this ensemble to one with fluctuating extensive variables because there is only one: M, which we fix (assuming N is constant), i.e., we can't have E flowing independently of M from the system to an energy bath.

So we want to use E, H, and N as our natural variables.

In this ensemble,

$$M = \mu \sum_{i=1}^{N} n_i = \mu(n_+ - n_-) \qquad \text{where } n_+ = \text{number of spins up}$$
$$n_- = \text{number of spins down}$$

$$= \mu(2n_+ - N) , \qquad \text{and } n_+ + n_- = N$$

and

$$E = -H\mu(2n_+ - N) .$$

Since

$$\Omega(E) = \frac{N!}{n_+!(N-n_+)!} \qquad \text{and} \qquad S/k_B = \ln\Omega(E) ,$$

$$\beta = \frac{1}{k_B}\left.\frac{\partial S}{\partial E}\right|_{N,H} = \frac{\partial \ln\Omega(E)}{\partial E} = \frac{\partial \ln\Omega}{\partial n_+}\frac{\partial n_+}{\partial E} = -\frac{\partial}{\partial n_+}\left(\ln[n_+!(N-n_+)!]\right) \cdot \frac{1}{-2H\mu} .$$

So

$$\beta H = \frac{1}{2\mu}\frac{\partial}{\partial n_+}\ln[n_+!(N-n_+)!] .$$

For $N >> 0$,

$$\beta H = \frac{1}{2\mu}\frac{\partial}{\partial n_+}[n_+\ln n_+ + (N-n_+)\ln(N-n_+) - N]$$

$$= \frac{1}{2\mu}[\ln n_+ - \ln(N-n_+)] = \frac{1}{2\mu}\ln\left(\frac{n_+}{N-n_+}\right) .$$

Thus,

$$n_+ = \frac{N}{1 + e^{-2\beta\mu H}} ,$$

hence

$$M = \mu(2n_+ - N) = N\mu\tanh(\beta\mu H)$$

which agrees with Exercise 3.19 and (after a change in notation) with 3.9 as well.

Substituting $n_+ = M/2\mu + N/2$ into the above equation for βH gives

$$\beta H = \frac{1}{2\mu} \ln \left(\frac{N\mu + M}{N\mu - M}\right) .$$

3.21 In the $\{A,B\}$ basis, $H_0 = \begin{pmatrix} 0 & -\Delta \\ -\Delta & 0 \end{pmatrix}$ and $m = \begin{pmatrix} \mu & 0 \\ 0 & -\mu \end{pmatrix}$.

In Pauli spin matrices notation, $H_0 = -\Delta\sigma_x$ and $m = \mu\sigma_z$.

Then $H = H_0 - m\mathbf{E} = -\Delta\sigma_x - \mu\mathbf{E}\sigma_z$.

(a) $\mathbf{E} = 0 \qquad H = H_0 = -\Delta\sigma_x = -\Delta\begin{pmatrix} 0 & 1 \\ 1 & 0 \end{pmatrix}$

$$|\pm\rangle = \frac{1}{\sqrt{2}}[|A\rangle \pm |B\rangle] \text{ , i.e. } |+\rangle = \frac{1}{\sqrt{2}}\begin{pmatrix} 1 \\ +1 \end{pmatrix} \text{ and } |-\rangle = \frac{1}{\sqrt{2}}\begin{pmatrix} 1 \\ -1 \end{pmatrix} .$$

So

$$H|+\rangle = -\Delta/\sqrt{2}\begin{pmatrix} 0 & 1 \\ 1 & 0 \end{pmatrix}\begin{pmatrix} 1 \\ 1 \end{pmatrix} = -\Delta/\sqrt{2}\begin{pmatrix} 1 \\ 1 \end{pmatrix} = -\Delta|+\rangle \; ; \quad E_+ = -\Delta ,$$

$$H|-\rangle = -\Delta/\sqrt{2}\begin{pmatrix} 0 & 1 \\ 1 & 0 \end{pmatrix}\begin{pmatrix} 1 \\ -1 \end{pmatrix} = -\Delta/\sqrt{2}\begin{pmatrix} -1 \\ 1 \end{pmatrix} = \Delta|-\rangle \; ; \quad E_- = +\Delta .$$

(b) H in the $\{+,-\}$ basis is $-\Delta\begin{pmatrix} 1 & 0 \\ 0 & -1 \end{pmatrix}$.

Since H is diagonal, $e^{-\beta H} = \begin{pmatrix} e^{\beta\Delta} & 0 \\ 0 & e^{-\beta\Delta} \end{pmatrix}$, and

$$Q = \mathrm{Tr}e^{-\beta H} = \mathrm{Tr}\begin{pmatrix} e^{\beta\Delta} & 0 \\ 0 & e^{-\beta\Delta} \end{pmatrix} = e^{\beta\Delta} + e^{-\beta\Delta} \quad \text{as expected.}$$

Doing the trace with $|A\rangle$ and $|B\rangle$ gives

$$Q = \langle A|\begin{pmatrix} e^{\beta\Delta} & 0 \\ 0 & e^{-\beta\Delta} \end{pmatrix}|A\rangle + \langle B|\begin{pmatrix} e^{\beta\Delta} & 0 \\ 0 & e^{-\beta\Delta} \end{pmatrix}|B\rangle \qquad \begin{array}{l} |A\rangle = [|+\rangle + |-\rangle]/\sqrt{2} \\ |B\rangle = [|+\rangle - |-\rangle]/\sqrt{2} \end{array}$$

$$= \frac{1}{2}[e^{\beta\Delta} + e^{-\beta\Delta}] + \frac{1}{2}[e^{\beta\Delta} + e^{-\beta\Delta}]$$

$$= e^{\beta\Delta} + e^{-\beta\Delta} .$$

In the $\{A,B\}$ basis $\quad H = -\Delta \begin{pmatrix} 0 & 1 \\ 1 & 0 \end{pmatrix}$,

and $$e^{-\beta H} = e^{\beta\Delta \begin{pmatrix} 0 & 1 \\ 1 & 0 \end{pmatrix}} = \sum_{n=0}^{\infty} \frac{[\beta\Delta \begin{pmatrix} 0 & 1 \\ 1 & 0 \end{pmatrix}]^n}{n!}$$

$$= \sum_{\substack{n=0 \\ \text{even}}}^{\infty} \frac{(\beta\Delta)^n}{n!} \begin{pmatrix} 1 & 0 \\ 0 & 1 \end{pmatrix} + \sum_{\substack{n=0 \\ \text{odd}}}^{\infty} \frac{(\beta\Delta)^n}{n!} \begin{pmatrix} 0 & 1 \\ 1 & 0 \end{pmatrix}$$

$$= \frac{1}{2} [\sum_{n=0}^{\infty} \frac{(\beta\Delta)^n}{n!} + \sum_{n=0}^{\infty} \frac{(-\beta\Delta)^n}{n!}] \begin{pmatrix} 1 & 0 \\ 0 & 1 \end{pmatrix} + \frac{1}{2} [\sum_{n=0}^{\infty} \frac{(\beta\Delta)^n}{n!} - \sum_{n=0}^{\infty} \frac{(-\beta\Delta)^n}{n!}] \begin{pmatrix} 0 & 1 \\ 1 & 0 \end{pmatrix}$$

$$= \frac{1}{2} [e^{\beta\Delta} + e^{-\beta\Delta}] \begin{pmatrix} 1 & 0 \\ 0 & 1 \end{pmatrix} + \frac{1}{2} [e^{\beta\Delta} - e^{-\beta\Delta}] \begin{pmatrix} 0 & 1 \\ 1 & 0 \end{pmatrix}$$

$$= \begin{pmatrix} \cosh \beta\Delta & \sinh \beta\Delta \\ \sinh \beta\Delta & \cosh \beta\Delta \end{pmatrix} .$$

These matrix elements are the same as those arrived at by rewriting $\begin{pmatrix} e^{\beta\Delta} & 0 \\ 0 & e^{-\beta\Delta} \end{pmatrix}$ in the $\{A,B\}$ basis. Now performing the trace,

$$\mathrm{Tr}\, e^{-\beta H} = 2 \cosh\beta\Delta = e^{\beta\Delta} + e^{-\beta\Delta} .$$

In the end, the route to the partition function is unimportant, since the trace is invariant to a change of basis.

(c) Using the $\{+, -\}$ basis,

(i) $$\langle m \rangle = \frac{1}{Q} \mathrm{Tr}(m e^{-\beta H}) = \frac{1}{Q}[\langle + | m | + \rangle e^{-\beta\Delta} + \langle - | m | - \rangle e^{+\beta\Delta}] .$$

Since

$$\langle + | m | + \rangle = \frac{1}{2} (\langle A | + \langle B |)\ m\ (| A \rangle + | B \rangle) = 0$$

and similarly $\langle - | m | - \rangle = 0$, therefore

$$\langle m \rangle = 0 .$$

(ii) Similarly $\langle + |\ |m|\ | + \rangle = e^{-\beta\Delta}$ and $\langle - |\ |m|\ | - \rangle = e^{\beta\Delta}$.

So $\langle|m|\rangle = |\mu|$.

(iii) $\langle(\delta m)^2\rangle = \langle m^2\rangle - \langle m\rangle^2 = \mu^2$.

(d) For $\mathbf{E} \neq 0$, $H = -\Delta\sigma_x - \mu\mathbf{E}\sigma_z = \begin{pmatrix} -\mu\mathbf{E} & -\Delta \\ -\Delta & +\mu\mathbf{E} \end{pmatrix}$.

(i) The energies then satisfy the equation

$$\det(E\underset{\sim}{I} - \underset{\sim}{H}) = 0 .$$

So $E^2 - (\mu\mathbf{E})^2 - \Delta^2 = 0$, or $E = \pm\sqrt{(\mu\mathbf{E})^2 + \Delta^2}$

and $Q = e^{\beta\sqrt{(\mu\mathbf{E})^2 + \Delta^2}} + e^{-\beta\sqrt{(\mu\mathbf{E})^2 + \Delta^2}}$.

Then $A(\mathbf{E}) = -\beta^{-1} \ln[2\cosh(\beta\sqrt{(\mu\mathbf{E})^2 + \Delta^2})]$

and the free energy of solvation is

$$A(\mathbf{E}) - A(0) = -\beta^{-1}\left\{ \ln[2\cosh\left(\beta\sqrt{(\mu\mathbf{E})^2 + \Delta^2}\right)] - \ln[2\cosh(\beta\Delta)] \right\} .$$

(ii) $e^{-\beta H} = \sum_{n=0}^{\infty} \frac{[c\sigma_z + d\sigma_x]^n}{n!}$ where $c = \beta\mu\mathbf{E}$ and $d = \beta\Delta$.

To evaluate the $n^{\underline{th}}$ term, note

$$[c\sigma_z + d\sigma_x]^2 = c^2 I + d^2 I + cd(\sigma_z\sigma_x + \sigma_x\sigma_z)$$

$$= (c^2 + d^2)I \qquad \text{since } \{\sigma_z, \sigma_x\} = 0 .$$

So $[c\sigma_z + d\sigma_x]^n = (c^2 + d^2)^{n/2}\, I$ for n even

$= (c^2 + d^2)^{(n-1)/2}\, [c\sigma_z + d\sigma_x]$ for n odd

and

$$e^{-\beta H} = \sum_{\substack{n=0 \\ \text{even}}}^{\infty} \frac{(\sqrt{c^2 + d^2})^n\, I}{n!}$$

$$+ \sum_{\substack{n=0 \\ n\ \text{odd}}}^{\infty} \frac{\sqrt{(c^2 + d^2)^n}}{n!} \left(\frac{c}{\sqrt{c^2 + d^2}}\,\sigma_z + \frac{d}{\sqrt{c^2 + d^2}}\,\sigma_x\right)$$

$$= \cosh(\sqrt{c^2+d^2}) \cdot I + \sinh(\sqrt{c^2+d^2}) \left(\frac{c}{\sqrt{c^2+d^2}} \sigma_z + \frac{d}{\sqrt{c^2+d^2}} \sigma_x\right) .$$

Since TrI = 2, and $\mathrm{Tr}\,\sigma_z = \mathrm{Tr}\,\sigma_x = 0$,

$$Q = 2 \cosh(\beta\sqrt{\mu^2\mathbf{E}^2 + \Delta^2})$$ in agreement with (i).

(e) To evaluate $\langle m\rangle = \mathrm{Tr}(me^{-\beta H})/\mathrm{Tr}\, e^{-\beta H}$

we want the matrix elements of m in the H eigenstate basis $|\pm_{\mathbf{E}}\rangle$.

Solving for $|\pm\rangle$ in the $\{A,B\}$ basis,

$$\begin{pmatrix} -\mu\mathbf{E} & -\Delta \\ -\Delta & \mu\mathbf{E} \end{pmatrix} \begin{pmatrix} x \\ y \end{pmatrix} = \pm \sqrt{(\mu\mathbf{E})^2 + \Delta^2} \begin{pmatrix} x \\ y \end{pmatrix}$$ implies

$$|+\rangle = \begin{pmatrix} \Delta \\ -\mu\mathbf{E} + E \end{pmatrix} / \sqrt{\Delta^2 + (E-\mu\mathbf{E})^2}$$ having energy $-\sqrt{(\mu\mathbf{E})^2 + \Delta^2} = -E$

$$|-\rangle = \begin{pmatrix} \Delta \\ -\mu\mathbf{E} - E \end{pmatrix} / \sqrt{\Delta^2 + (E-\mu\mathbf{E})^2}$$ having energy $\sqrt{(\mu\mathbf{E})^2 + \Delta^2} = E$.

So $$\langle\pm |m| \pm\rangle = \mu \frac{\Delta^2 - (E \mp \mu\mathbf{E})^2}{\Delta^2 + (E \mp \mu\mathbf{E})^2} = \mu \frac{\pm\mu\mathbf{E}}{E} .$$

Therefore $$\langle m\rangle = \frac{\mu^2\mathbf{E}}{\sqrt{(\mu\mathbf{E})^2 + \Delta^2}} \{ \exp[\beta\sqrt{\mu^2\mathbf{E}^2 + \Delta^2}] - \exp[-\beta\sqrt{\mu^2\mathbf{E}^2 + \Delta^2}] \}/Q$$

$$= \frac{\mu^2\mathbf{E}}{\sqrt{\mu^2\mathbf{E}^2 + \Delta^2}} \tanh[\beta\sqrt{\mu^2\mathbf{E}^2 + \Delta^2}] .$$

Similarly $\langle \pm |m| \pm \rangle = |\mu| \dfrac{\Delta^2 + (E \mp \mu\mathbf{E})^2}{\Delta^2 + (E \mp \mu\mathbf{E})^2}$, hence $\langle |m| \rangle = |\mu|$. This is true for any E since $\langle |m| \rangle = \mathrm{Tr}(|\mu| I\, e^{-\beta H})/\mathrm{Tr}\, e^{-\beta H}$.

In other words, $\langle m\rangle$ increases as the field increases. There is a competition between the field **E** which would make $|A\rangle$ and $|B\rangle$ the eigenstates, and the tunneling Δ which has $|\pm\rangle$ as eigenstates. Only as $|\mathbf{E}| \to \infty$, does $\langle m\rangle_{\mathbf{E}\to\infty} = \mu$ and $\langle m\rangle_{\mathbf{E}\to -\infty} = -\mu$, where all spins are aligned with (or against) the field.

As a final note, although $\langle m\rangle = -\frac{\partial \ell nQ}{\partial \beta \mathbf{E}}$ for small **E**, the fluctuation in m, $\langle(\delta m)^2\rangle$, is in general no longer equal to $-\frac{\partial\langle m\rangle}{\partial \beta \mathbf{E}}$ unless m commutes with H. Systematic analysis of the fluctuation in that case is in fact equivalent to analyzing the quantum dynamics of the system, which is not the focus of this text.

3.22 $\beta p = \rho/(1-b\rho) - \beta a\rho^2 , \qquad \rho = \frac{\langle N\rangle}{V} , \qquad V = L^3 .$

(a) For constant $V = L^3$ $\quad \langle(\delta\rho)^2\rangle^{1/2}/\rho = \langle(\delta N)^2\rangle^{1/2}/\langle N\rangle$.

$$\langle(\delta N)^2\rangle = \left(\frac{\partial\langle N\rangle}{\partial(\beta\mu)}\right)_{\beta,V} = V\left(\frac{\partial\rho}{\partial\beta\mu}\right)_{\beta,V} = V\left(\frac{\partial\rho}{\partial\beta p}\right)_{\beta,V}$$

$$= \langle N\rangle \left(\frac{\partial\beta p}{\partial\rho}\right)^{-1}_{\beta,V}$$

$$= \langle N\rangle \frac{(1-b\rho)^2}{1-2\beta a\rho(1-b\rho)^2} ,$$

or

$$\frac{\langle(\delta\rho)^2\rangle^{1/2}}{\rho} = \frac{1}{\sqrt{\rho V}} \cdot \frac{(1-b\rho)}{\sqrt{1-2\beta a\rho(1-b\rho)^2}}$$

which vanishes as the volume becomes infinite.

(b) At the critical point

$$\left(\frac{\partial\beta p}{\partial\rho}\right)_\beta = \frac{1}{(1-b\rho)^2} - 2\beta a\rho = 0 \quad \text{and} \quad \left(\frac{\partial^2\beta p}{\partial\rho^2}\right)_\beta = \frac{2b}{(1-b\rho)^3} - 2\beta a = 0 .$$

Do some algebra!

$$\rho_c = \frac{1}{3b} \quad \text{and} \quad \beta_c = \frac{1}{2a\rho_c(1-b\rho_c)^2} = \frac{3b}{2a(2/3)^2} = \frac{27b}{8a} .$$

(c) Use the result of Part (a) with $V = 100b$, $\rho = \frac{1}{3b}$ and $\beta = \frac{1}{x}\frac{27b}{8a}$, where x will be close to 1:

$$\frac{\langle(\delta\rho)^2\rangle^{1/2}}{\rho} = \frac{1}{5\sqrt{3(1-1/x)}} .$$

Thus we have the table

x	$\langle(\delta\rho)^2\rangle^{1/2}/\rho$
1.1	0.38
1.001	3.7
1.00001	37

(d) We're looking for "significant" density fluctuations in a volume $V = \sim (1000Å)^3$. Typical molecular volumes are $\sim (5Å)^3 = b$
$V = (200)^3 b = 8 \times 10^6 b$.
To estimate significant $\frac{\langle(\delta\rho)^2\rangle^{1/2}}{\rho}$, we checked the CRC for the index of refraction of distinguishable (by sight) fluids:

$H_2O \approx 1.33$

n-Hexane ≈ 1.37 or $\frac{\Delta n}{n} \approx 3\%$.

For small differences in ρ, we can consider $n \propto \rho$. So take significant $\frac{\langle(\delta\rho)^2\rangle^{1/2}}{\rho} = 0.03$. Solving for the difference in $\beta = \beta_c/x$ at $\rho = \rho_c = \frac{1}{3b}$, we get:

$$\sqrt{1 - 1/x} = 10^{-3} \frac{1}{\sqrt{6}} \left(\frac{\langle(\delta\rho)^2\rangle^{1/2}}{\rho}\right)^{-1} .$$

$$\text{Then } x \approx 1 + 10^{-5} \underbrace{\left(\frac{\langle(\delta\rho)^2\rangle^{1/2}}{\rho}\right)^{-1}}_{\approx 10^{-2}} \approx 1 + 10^{-3}$$

In other words, critical fluctuations will be easily observed optically if you are within 0.1% of the critical temperature.

3.23 $A \rightleftarrows B$, $N_A + N_B = N$.

We can make an analogy with the ideal gas analysis of Sec. 3.6.

$$\langle(N_A - \langle N_A\rangle)^2\rangle = \langle N_A^2\rangle - \langle N_A\rangle^2$$

$$= \sum_{ij}^{N} [\langle n_{ai} n_{aj}\rangle - \langle n_{ai}\rangle\langle n_{aj}\rangle] , \quad n_{ai} = 1 , \text{ if molecule } i \text{ is in state A}$$
$$= 0 , \text{ otherwise}$$

$$= \sum_{i}^{N} [\langle n_{ai}^2 \rangle - \langle n_{ai} \rangle^2] + \underbrace{\sum_{i \neq j}^{N} [\langle n_{ai} n_{aj} \rangle - \langle n_{ai} \rangle \langle n_{aj} \rangle]}_{\text{zero since different molecules are uncorrelated}}$$

$$= N\langle n_{a1} \rangle (1 - \langle n_{a1} \rangle) \quad , \text{ since each molecule is equivalent}$$

$$= N\langle n_{a1} \rangle \langle 1 - n_{a1} \rangle = x_A x_B N \ ,$$

where

$x_A = \langle N_A \rangle / N$, and $x_B = 1 - x_A$.

4.5 Begin with the equation at the top of page 94,

$$\beta A = -\ln Q = \int_0^\infty d\omega\; g(\omega)\; \ln\,[e^{\beta\hbar\omega/2} - e^{-\beta\hbar\omega/2}]$$

$$= (\frac{ND^2}{\omega_0^D}) \int_0^{\omega_0} d\omega\; \omega^{D-1}\; \ln\,[e^{\beta\hbar\omega/2} - e^{-\beta\hbar\omega/2}] \;.$$

As long as $\beta\hbar\omega_0/2 >> 1$, we can approximate this integral in the limit of $\beta\to\infty$ as

$$\lim_{\substack{T\to 0\\(\beta\to\infty)}} \beta A = (\frac{ND^2}{\omega_0^D}) \int_0^{\omega_0} d\omega\; \omega^{D-1}\; \beta\hbar\omega/2$$

$$= \frac{ND^2\omega_0\hbar}{2(D+1)}\;\beta \;.$$

Hence

$$\langle E\rangle = \frac{\partial \ln Q}{\partial(-\beta)}\Big|_{N,V} \sim \frac{ND^2\omega_0\hbar}{2(D+1)} \;.$$

Similarly,

$$C_V = (\frac{\partial E}{\partial T})_V = -\,k_B\beta^2 \frac{\partial}{\partial\beta}\,(\frac{\partial \ln Q}{\partial(-\beta)})_V$$

$$= -\,k_B\beta^2 \frac{\partial}{\partial\beta}\,[\int_0^\infty d\omega\; g(\omega)\; \frac{e^{\beta\hbar\omega/2} + e^{-\beta\hbar\omega/2}}{e^{\beta\hbar\omega/2} - e^{-\beta\hbar\omega/2}}\;\hbar\omega/2\;]$$

$$= -\,k_B\beta^2\,(\frac{ND^2}{\omega_0^D}) \int_0^{\omega_0} d\omega\; \omega^{D-1}\; (\hbar\omega/2)^2\;[-(\frac{2}{e^{\beta\hbar\omega/2} - e^{-\beta\hbar\omega/2}})^2] \;.$$

Change variables to $x = \beta\hbar\omega$ and let $k_B\Theta_D \equiv \hbar\omega_0$:

$$C_V = k_B\,\frac{ND^2}{k_B^D\,\Theta^D}\,\hbar \int_0^{\Theta_D/T} \frac{dx}{\beta\hbar}\;\frac{x^{D-1}}{\beta^{D-1}}\;\frac{x^2 e^x}{(e^x-1)^2}$$

$$= k_B ND^2\,(\frac{T}{\Theta_D})^D \int_0^{\Theta_D/T} \frac{x^{D+1}e^x}{(e^x-1)^2}\;dx \;.$$

In the limit $T \to 0$, we encounter the integral

$$\int_0^\infty \frac{x^{D+1}e^x}{(e^x-1)^2}\;dx = x^{D+1}\,\frac{-1}{e^x-1}\Big|_0^\infty + (D+1)\int_0^\infty \frac{x^D}{e^x-1}\;dx$$

$$= 0 + (D+1)\ \underbrace{\Gamma(D+1)}_{\text{Gamma Function}}\ \underbrace{\zeta(D+1)}_{\text{Reimann Zeta Function}} = (D+1)!\ \zeta(D+1) .$$

For D = 3, we note $\zeta(4) = \pi^4/90$, and

$$C_V \sim \frac{12}{5}\pi^4 Nk_B \left(\frac{T}{\Theta_D}\right)^3 .$$

4.7 $$\langle n_i n_j\rangle = \frac{1}{\Xi}\frac{\partial^2 \Xi}{\partial\beta\varepsilon_i \partial\beta\varepsilon_j} = \frac{1}{\Xi}\frac{\partial}{\partial(-\beta\varepsilon_j)}(\Xi \langle n_i\rangle)$$

$$= \langle n_i\rangle \langle n_j\rangle \qquad i \neq j , \qquad \text{since } n_i \text{ and } n_j \text{ are uncorrelated}$$

$$= \langle n_i\rangle \qquad i = j , \qquad \text{since } \langle n_i^2\rangle = \langle n_i\rangle .$$

Therefore,

$$g_{ij} = \langle n_i\rangle\langle n_j\rangle (1-\delta_{ij}) .$$

4.11 The expression for the electronic partition function of an H atom, $q_{int}(T) = g_1 e^{\beta\varepsilon_0} + g_2 e^{\beta\varepsilon_0/4} + \ldots + g_n e^{\beta\varepsilon_0/n^2} + \ldots$ is only exact in the case of an isolated atom. When the atom is put in a box or if there are other atoms in the universe, the convergence of energy levels $\rightarrow E = 0$ as $n \rightarrow \infty$ is lifted. The reason for this is that as n increases, so does the energy and spatial extent of the atom. The expectation value for the repulsive energy of two atoms then has no upper bound, and will increase as the spatial overlap increases.

4.15 For an ideal mixture

$$Q = \frac{q_A^{N_A}}{N_A!}\ \frac{q_B^{N_B}}{N_B!} ,$$

Further

$$-\beta\mu_A = \left(\frac{\partial \ln Q}{\partial N_A}\right)_{\beta,V,N_B} , \quad -\beta\mu_B = \left(\frac{\partial \ln Q}{\partial N_B}\right)_{\beta,V,N_A} .$$

Using $\ell n N! = N \ell n N - N$ we get

$$-\ell n Q = - N_A \ell n\, q_A + \ell n\, N_A! - N_B \ell n\, q_B + \ell n\, N_B! \ ,$$

so that

$$\mu_A = \mu_B \ \text{ implies } \ -\ell n\, q_A + \ell n\, N_A = -\ell n\, q_B + \ell n\, N_B \ ,$$

or

$$\frac{N_A}{N_B} = \frac{q_A}{q_B} = \frac{g_A}{g_B}\, e^{-\beta\Delta\varepsilon} \ .$$

4.16 (a) $Q = \frac{1}{N!}\, q^N = \frac{1}{N!}\,(q_A + q_B)^N$

$$= \frac{1}{N!} \sum_{N_A=0}^{N} \frac{N!}{N_A!(N-N_A)!}\, q_A^{N_A}\, q_B^{(N-N_A)}$$

$$= \underbrace{\sum_{N_A, N_B}}_{\substack{\text{such that} \\ N_A + N_B = N}} \frac{1}{N_A!N_B!}\, q_A^{N_A}\, q_B^{N_B} \equiv \sum_P \exp[-\beta A(N_A, N_B)]$$

(b) $\mu_A = \mu_B$ implies $\dfrac{\partial A}{\partial \langle N_A\rangle} = \dfrac{\partial A}{\partial \langle N_B\rangle}$. (1)

$\langle N_A\rangle + \langle N_B\rangle = N = \text{const}$ implies $\dfrac{\partial A}{\partial \langle N_A\rangle} = -\dfrac{\partial A}{\partial \langle N_B\rangle}$. (2)

Equations (1) and (2) imply

$$-\frac{\partial A}{\partial \langle N_A\rangle} = \frac{\partial A}{\partial \langle N_A\rangle} = 0 = \frac{\partial A}{\partial \langle N_B\rangle} \ .$$

Therefore, quite generally, $A(N_A,N_B)$, is minimized when $N_A + N_B = N$ are partitioned to satisfy chemical equilibrium. More specifically, with the ideal expression in Part (a), we have

$$0 = \frac{\partial}{\partial N_A} \{- \ell n[N_A!(N-N_A)!] + N_A\, \ell n\, q_A + (N-N_A)\ell n\, q_B\}$$

$$= - \ell n[N_A/(N-N_A)] + \ell n(q_A/q_B) \ ,$$ applying Stirling's approximation

or associating the solution to this equation with

$\langle N_A \rangle = N - \langle N_B \rangle$, we obtain

$\langle N_A \rangle / \langle N_B \rangle = q_A / q_B$.

4.17 $$Q = \sum_{\substack{N_A, N_B \\ (N_A + N_B = N)}} \frac{q_A^{N_A}}{N_A!} \frac{q_B^{N_B}}{N_B!} = (q_A + q_B)^N / N! \quad \text{as shown above .}$$

$$\langle N_A \rangle = \frac{1}{Q} \sum_{\substack{N_A, N_B \\ (N_A + N_B = N)}} N_A \frac{q_A^{N_A}}{N_A!} \frac{q_B^{N_B}}{N_B!} = \frac{q_A}{Q} \sum_{\substack{N_A, N_B \\ (N_A + N_B = N)}} N_A \frac{q_A^{N_A - 1}}{N_A!} \frac{q_B^{N_B}}{N_B!}$$

$$= q_A \left(\frac{\partial \ln Q}{\partial q_A}\right)_{q_B, N}$$

$$= q_A \frac{\partial}{\partial q_A} \left[\ln \left(\frac{(q_A + q_B)^N}{N!}\right)\right]_{q_B, N} = q_A \frac{N}{(q_A + q_B)} .$$

Since A and B play symmetrical roles, $\langle N_B \rangle = q_B \dfrac{N}{(q_A + q_B)}$.

Hence,

$\langle N_A \rangle / \langle N_B \rangle = q_A / q_B$.

Similarly,

$$\langle (N_A - \langle N_A \rangle)^2 \rangle = \langle N_A^2 \rangle - \langle N_A \rangle^2 = \frac{1}{Q} \sum_{\substack{N_A, N_B \\ (N_A + N_B = N)}} N_A^2 \frac{q_A^{N_A} q_B^{N_B}}{N_A! N_B!} - \frac{q_A^2}{Q^2} \left(\frac{\partial Q}{\partial q_A}\right)^2_{q_B, N}$$

$$= \frac{q_A}{Q} \frac{\partial}{\partial q_A} \left(q_A \frac{\partial Q}{\partial q_A}\right)_{q_B, N} - \frac{q_A^2}{Q^2} \left(\frac{\partial Q}{\partial q_A}\right)^2_{q_B, N} = q_A \left[\frac{q_A}{Q} \frac{\partial^2 Q}{\partial q_A^2} + \frac{1}{Q} \frac{\partial Q}{\partial q_A} - \frac{q_A}{Q^2} \left(\frac{\partial Q}{\partial q_A}\right)^2\right]$$

$$= q_A \frac{\partial}{\partial q_A} \left[\frac{q_A}{Q} \frac{\partial Q}{\partial q_A}\right]_{q_B, N} = q_A \left.\frac{\partial \langle N_A \rangle}{\partial q_A}\right|_{q_B, N} = q_A \left[\frac{N}{q_A + q_B} - q_A \frac{N}{(q_A + q_B)^2}\right]$$

$$= q_A N \left[q_B / (q_A + q_B)^2\right] = \langle N_A \rangle \langle N_B \rangle / N .$$

4.18 For an ideal gas of structureless particles

$$\varepsilon_{\underset{\sim}{k}} = \frac{\hbar^2 k^2}{2m} \; ; \quad \underset{\sim}{k} = \frac{\pi}{L}(n_x\hat{x} + n_y\hat{y} + n_z\hat{z}) \quad .$$

(a) Then assuming a macroscopic volume,

$$\beta pV = \ell n\Xi = \sum_{\underset{\sim}{k}} \ell n[1 + e^{\beta(\mu-\varepsilon_{\underset{\sim}{k}})}] = \frac{V}{(2\pi)^3}\int_0^\infty 4\pi k^2 \ell n[1 + e^{\beta(\mu-\hbar^2k^2/2m)}]dk$$

$$= \frac{4V}{\pi^2}\left(\frac{m}{2\beta\hbar^2}\right)^{3/2}\int_0^\infty x^2\, \ell n(1 + ze^{-x^2})\, dx \,. \qquad z = e^{\beta\mu}$$

So $\beta p = \frac{4}{\sqrt{\pi}}\,\frac{1}{\lambda^3}\int_0^\infty x^2\, \ell n(1 + ze^{-x^2})\, dx \qquad \lambda = \left(\frac{2\pi\beta\hbar^2}{m}\right)^{1/2}$

$$= \frac{1}{\lambda^3}\, f_{5/2}(z) \,.$$

Substituting the series for $\ell n(1+x)$,

$$f_{5/2}(z) = \frac{4}{\sqrt{\pi}}\sum_{k=1}^\infty (-1)^{k+1}\frac{z^k}{k}\int_0^\infty x^2 e^{-kx^2} dx$$

$$= \sum_{k=1}^\infty (-1)^{k+1}\, z^k/\, k^{5/2} \,.$$

To calculate the density,

$$\langle N\rangle = \sum_{\underset{\sim}{k}} \langle n_{\underset{\sim}{k}}\rangle = \sum_{\underset{\sim}{k}} \frac{1}{e^{\beta(\varepsilon_{\underset{\sim}{k}}-\mu)} + 1} = \frac{V}{(2\pi)^3}\int_0^\infty 4\pi k^2\, \frac{ze^{-\hbar^2k^2/2m}}{1 + ze^{-\hbar^2k^2/2m}}\, dk$$

$$= \frac{V}{\lambda^3}\,\frac{4}{\sqrt{\pi}}\int_0^\infty x^2\,\frac{ze^{-x^2}}{1 + ze^{-x^2}}\, dx \,.$$

So

$$\rho\lambda^3 = \frac{4}{\sqrt{\pi}}\sum_{k=1}^\infty (-1)^{k+1} z^k \int_0^\infty x^2\, e^{-kx^2}\, dx$$

$$= \sum_{k=1}^\infty (-1)^{k+1}\, z^k/\, k^{3/2}$$

$$= f_{3/2}(z) \,.$$

(b) $$\langle E\rangle = \sum_{\underset{\sim}{k}} \langle n_{\underset{\sim}{k}}\rangle \varepsilon_k = \frac{V}{(2\pi)^3}\int_0^\infty 4\pi k^2 \left(\frac{\hbar^2k^2}{2m}\right)\cdot \frac{1}{1+e^{\beta(\hbar^2k^2/2m - \mu)}}\, dk$$

$$= \frac{m}{\beta}\frac{\partial}{\partial m}\frac{V}{(2\pi)^3}\int_0^\infty 4\pi k^2 \,\ln\left[1+e^{\beta(\mu - \hbar^2k^2/2m)}\right] dk$$

$$= \frac{m}{\beta}\frac{\partial}{\partial m}\frac{V}{\lambda^3} f_{5/2}(z) = \frac{m}{\beta}\;\frac{3}{2m}\;\frac{V}{\lambda^3} f_{5/2}(z)$$

$$= \frac{3}{2\beta}(\beta pV) = \frac{3}{2}\, pV\ .$$

4.19 (a) We want to invert $y = \rho\lambda^3 = \sum_{\ell=1}^{\infty} (-1)^{\ell+1} z^\ell/\ell^{3/2}$.

Using $z = \sum_{k=0}^{\infty} c_k y^k$ in the method of undetermined coefficients,

$$y = \sum_{\ell=1}^{\infty}(-1)^{\ell+1}\left(\sum_{k=0}^{\infty} c_k y^k\right)^\ell/\ \ell^{3/2}$$

$$= \sum_{\ell=1}^{\infty}\frac{(-1)^{\ell+1}}{\ell^{3/2}}\, c_0^\ell + \left[\sum_{\ell=1}^{\infty}\frac{(-1)^{\ell+1}}{\ell^{3/2}}\,\ell\, c_0^{\ell-1}\, c_1\right] y$$

$$+ \left[\sum_{\ell=1}^{\infty}\frac{(-1)^{\ell+1}}{\ell^{3/2}}\,\ell\, c_0^{\ell-1}\, c_2 + \sum_{\ell=2}^{\infty}\frac{(-1)^{\ell+1}}{\ell^{3/2}}\binom{\ell}{2} c_0^{\ell-2}\, c_1^2\right] y^2 + \dots$$

Equating terms,

$$0 = f_{3/2}(c_0) \quad \text{which implies} \quad c_0 = 0$$

and then

$$1 = c_1$$

and

$$0 = c_2 - \frac{1}{2^{3/2}}\, c_1^2\ .$$

Therefore $\quad z = \rho\lambda^3 + (\rho\lambda^3)^2/2\sqrt{2} + \dots\ .$

(b) $$\langle n_{\underset{\sim}{p}}\rangle = \frac{ze^{-\beta p^2/2m}}{1+ze^{-\beta p^2/2m}} \approx \rho\lambda^3 e^{-\beta p^2/2m} \qquad \text{for } \rho\lambda^3 \ll 1\ .$$

(c) $$\langle |p|\rangle = \frac{V}{(2\pi)^3}\,\hbar^{-3}\int_0^\infty 4\pi p^3\, \rho\lambda^3\, e^{-\beta p^2/2m}\, dp$$

$$= \frac{V\rho}{2\pi^2\hbar^3}\,\lambda^3\,\frac{2m^2}{\beta^2}\ .$$

For a single particle $V\rho = 1$, so

$$\langle|p|\rangle = \left(\frac{2}{\pi}\right) h/\lambda \quad , \quad \text{or} \quad \lambda = \left(\frac{2}{\pi}\right) h/\langle|p|\rangle \quad .$$

(d)

$$\begin{aligned}\beta p/\rho &= \frac{1}{\rho\lambda^3} f_{5/2}(z) \\ &= \frac{1}{\rho\lambda^3} \{[\rho\lambda^3 + (\rho\lambda^3)^2/2^{3/2}] - (\rho\lambda^3)^2/ 2^{5/2} + \ldots\} \\ &= 1 + \rho\lambda^3/2^{5/2} + \ldots\end{aligned}$$

The finite $\rho\lambda^3$ correction is due to the negative correlation between fermions (i.e., the Pauli exclusion principle implies that no two identical particles can simultaneously exist at the same point in space). However, as $\rho\lambda^3 \rightarrow 0$, the effect of the exclusion principle diminishes as the particles become farther apart.

4.20 (a)

$$\begin{aligned}f_{3/2}(z) &= 4\,\pi^{-1/2} \int_0^\infty dx\, x^2[z^{-1}e^{x^2} + 1]^{-1} \\ &= 2\pi^{-1/2} \int_0^\infty dy\sqrt{y}\, [z^{-1}e^{y} + 1]^{-1}\end{aligned}$$

Since the derivative of the distribution $[z^{-1}e^{y} + 1]^{-1}$ is sharply peaked at $y = \ell nz$, we integrate by parts to get

$$f_{3/2}(z) = \frac{4}{3\sqrt{\pi}} \int_0^\infty dy\, y^{3/2}\, e^{y-\ell nz}[e^{y-\ell nz} + 1]^{-2} \, .$$

Expanding $y^{3/2}$ around the peak at $y = \ell nz$ gives

$$f_{3/2}(z) = \frac{4}{3\sqrt{\pi}} \int_{-\ell nz}^\infty dt\, \frac{e^t}{(e^t+1)^2} [(\ell nz)^{3/2} + 3/2(\ell nz)^{1/2}t + 3/8(\ell nz)^{-1/2}t^2 + \ldots]$$

$$\approx \frac{4}{3\sqrt{\pi}} (\ell nz)^{3/2} \int_{-\ell nz}^\infty dt\, \frac{e^t}{(e^t+ 1)^2}$$

$$\approx \frac{4}{3\sqrt{\pi}} (\ell nz)^{3/2} \left[\int_{-\infty}^\infty dt\, \frac{e^t}{(e^t + 1)^2} - \int_{-\infty}^{-\ell nz} dt\, \frac{e^t}{(e^t + 1)^2} \right] .$$

The first integral is 1 and the second is of the order z^{-1}.

So $\rho\lambda^3 \approx (\ell nz)^{3/2}\ 4/3\sqrt{\pi}$

and $z \approx e^{\beta\varepsilon_F}$ where $\varepsilon_F = (\hbar^2/2m)(6\pi^2\rho)^{2/3}$.

(b) $$\beta p\lambda^3 = f_{5/2}(z) = \frac{4}{\sqrt{\pi}} \int_0^\infty dx\ x^2\ \ell n(1 + e^{\beta\varepsilon_F - x^2})$$

$$= \varepsilon_F \int_0^\beta d\beta'\ f_{3/2}(z)$$

To get the correction term in $(\beta\varepsilon_F)^{-2}$, we calculate $f_{3/2}(z)$ to order t^2:

$$\int_{-\infty}^{\infty} dt\ \frac{t\ e^t}{(e^t + 1)^2} = 0$$ since the integrand is odd, and

$$\int_{-\infty}^{\infty} dt\ \frac{t^2 e^t}{(e^t + 1)^2} = \frac{\pi^2}{3} .$$

So $$\beta p\lambda^3 = \varepsilon_F \int_0^\infty d\beta'\ \frac{4}{3\sqrt{\pi}} [(\beta'\varepsilon_F)^{3/2} + \frac{\pi^2}{8}(\beta'\varepsilon_F)^{-1/2} + \ldots]$$

$$= \frac{4}{3\sqrt{\pi}} (\beta\varepsilon_F)^{3/2} [\frac{2}{5}(\beta\varepsilon_F) + \frac{\pi^2}{4}(\beta\varepsilon_F)^{-1} + \ldots] .$$

Hence $p = \frac{2}{5}\rho\ \varepsilon_F[1 + \frac{5\pi^2}{8}(\beta\varepsilon_F)^{-2} + \ldots]$.

The pressure is not zero at T=0 because the exclusion principle requires all but one fermion to have non-zero momentum.

4.21 (a) Simply plug in the Fermi distribution given on page 96.

(b) Since $e^{-\beta(\varepsilon-\varepsilon_F)} = x << 1$,

then $\frac{1}{1+x} \approx 1 - x$.

So,

$$p = 2\rho_s - 2\rho_s[1 + e^{-\beta(\varepsilon-\varepsilon_F)}]^{-1} \approx 2\rho_s[1 - (1 - e^{\beta(\varepsilon-\varepsilon_F)})]$$

$$= 2\rho_s e^{-\beta(\varepsilon-\varepsilon_F)} .$$

Similarly,

$$n = \frac{2}{(2\pi)^3} \int d\underset{\sim}{k} \, [1 + e^{\beta(\varepsilon_k + \varepsilon_F)}]^{-1} \approx \frac{2}{(2\pi)^3} \int d\underset{\sim}{k} \, e^{-\beta(\varepsilon_k+\varepsilon_F)}$$

$$= \frac{2}{(2\pi)^3} \, e^{-\beta\varepsilon_F} \, 4\pi \int_0^\infty dk \, k^2 \exp(-\beta\hbar^2 k^2/2m_e)$$

$$= e^{-\beta\varepsilon_F} \, 2 \left(\frac{m_e}{2\pi\beta\hbar^2}\right)^{3/2}$$

$$= e^{-\beta\varepsilon_F} \, 2/\lambda^3 .$$

Therefore,

$$pn = \rho_s/\lambda^3 4e^{-\beta\varepsilon} .$$

4.22 $$\langle N\rangle_{ad} = \sum_{j \text{ adsorbed}} \langle n_j\rangle = \underbrace{\rho_s V}_{\text{degeneracy}} \underbrace{e^{-\beta(-\varepsilon-\mu)}}_{\text{classical limit}} = e^{\beta\mu}\rho_s V e^{\beta\varepsilon}$$

$$\rho_{ad} = \frac{\langle N\rangle_{ad}}{V} = e^{\beta\mu}\rho_s e^{\beta\varepsilon} .$$

Also $\beta\mu = \ln\rho_g\lambda^3$ for the classical gas, and μ has the same value as in the above equation for ρ_{ad}.

Hence,

$$\rho_g = e^{\beta\mu}/\lambda^3 ,$$

which implies

$$\rho_{ad}/\rho_g = \rho_s\lambda^3 e^{\beta\varepsilon} .$$

4.23 (a) $E = E_A + E_B + E_C$

then

$$Q = \sum_{ABC} e^{-\beta(E_A + E_B + E_C)} = (\sum_A e^{-\beta E_A}) (\sum_B e^{-\beta E_B}) (\sum_C e^{-\beta E_C}) = q_A \, q_B \, q_C ,$$

$$\langle E\rangle = \frac{\partial \ln Q}{\partial(-\beta)} = \frac{\partial \ln q_A}{\partial(-\beta)} + \frac{\partial \ln q_B}{\partial(-\beta)} + \frac{\partial \ln q_C}{\partial(-\beta)} \text{ , and}$$

$$C_V = -k_B\beta^2 \partial\langle E\rangle/\partial\beta = k_B\beta^2 [\frac{\partial^2 \ln q_A}{\partial\beta^2} + \frac{\partial^2 \ln q_B}{\partial\beta^2} + \frac{\partial^2 \ln q_C}{\partial\beta^2}]$$

$$= C_V^{(A)} + C_V^{(B)} + C_V^{(C)} \text{ .}$$

Let $E = E_A + E_0$, E_0 = zero point energy .

Then $Q = \sum_A e^{-\beta(E_A + E_0)} = e^{-\beta E_0} \sum_A e^{-\beta E_A}$,

$$\langle E\rangle = \frac{\partial \ln Q}{\partial(-\beta)} = \frac{\partial}{\partial(-\beta)} [-\beta E_0 + \ln \sum_A e^{-\beta E_A}] = E_0 + \sum_A E_A \, e^{-\beta E_A}/Q \text{ , and}$$

$$C_V = - k_B\beta^2(\frac{\partial\langle E\rangle}{\partial\beta}) = - k_B\beta^2 \frac{\partial}{\partial\beta} (\sum_A E_A \, e^{-\beta E_A}/Q) \text{ ,}$$

which is independent of E_0.

In other words, fluctuations produce C_V, and E_0 has nothing to do with fluctuations.

(b) $q_{elec} = g_0 + g_1 e^{-\beta\varepsilon_1} + g_2 e^{-\beta\varepsilon_2}$

where energies ε_1 and ε_2 are relative to ε_0 , our choice for the zero of energy.

$$C_V^{(elec)} = k_B T^2 \frac{\partial^2}{\partial\beta^2} \ln q_{elec}$$

$$= k_B T^2 \{ \frac{1}{q_{elec}} [g_1\varepsilon_1^2 e^{-\beta\varepsilon_1} + g_2\varepsilon_2^2 e^{-\beta\varepsilon_2}] - \frac{1}{q_{elec}^2} [g_1\varepsilon_1 e^{-\beta\varepsilon_1} + g_2\varepsilon_2 e^{-\beta\varepsilon_2}]^2 \} \text{ .}$$

(c) $Q = q_{trans} \cdot q_{elec} \cdot q_{nuc} \cdot q_{rot} \cdot q_{vib}/\sigma_{AB}$, so

$$C_V = C_V^{(trans)} + C_V^{(elec)} + C_V^{(nuc)} + C_V^{(rot)} + C_V^{(vib)} \text{ .}$$

Now

$$C_V^{(elec)} = k_B\beta^2[\langle\varepsilon^2\rangle - \langle\varepsilon\rangle^2]$$

where

$$\langle\varepsilon^2\rangle - \langle\varepsilon\rangle^2 = \left(\frac{g_1\varepsilon_1^2 e^{-\beta\varepsilon_1} + \ldots}{g_0 + \ldots}\right) - \left(\frac{g_1\varepsilon_1 e^{-\beta\varepsilon_1} + \ldots}{g_0 + \ldots}\right)^2$$

$$\tilde{<} \; e^{-50,000/300} < e^{-100} \; .$$

So

$C_V^{(elec)} < k_B\beta^2 e^{-100}$ at room temperature, which is negligible .

Because of the high energies of excited electronic states, the contribution to C_V of excited states is negligible. Neglecting the excited states completely is equivalent to assuming $C_V^{(elec)} = 0$, which is therefore a good approximation.

Also, the above calculation shows that the ground state degeneracy doesn't enter into the heat capacity.

(d) $S = \dfrac{\langle E\rangle - A}{T}$

$$q_{elec} = \underbrace{g_0 e^{-\beta\varepsilon_0}}_{} + \underbrace{g_1 e^{-\beta\varepsilon_1} + g_2 e^{-\beta\varepsilon_2} + \ldots.}_{< e^{-100} e^{-\beta\varepsilon_0}}$$

Ignoring these terms doesn't affect q_{elec} significantly and also won't affect either $\langle E\rangle = \partial \ln q/\partial(-\beta)$ or $A = -\ln q/\beta$.

On the other hand, g_0 will affect S because even though the average energy doesn't care about the degeneracy, $\langle E\rangle = g_0\varepsilon_0 e^{-\beta\varepsilon_0}/g_0 e^{-\beta\varepsilon_0} = \varepsilon_0$, the free energy <u>is</u> affected.

$$A_{elec} = -\frac{1}{\beta}\ln g_0 + \varepsilon_0 \; .$$

$$S_{elec} = \frac{\varepsilon_0 - \varepsilon_0 + k_B T \ln g_0}{T} = k_B \ln g_0 \; .$$

Thus the degeneracy accounts for an additive term to the entropy.

4.24 $Q = q_{trans}^N \, q_{int}^N / N!$

If we write

$$q_{int} = g_0 e^{-\beta\varepsilon_0}(2I_A + 1)(2I_B + 1)\, q_{rot} q_{vib}$$

$$q_{rot} = T/\Theta_{rot}$$

$$q_{vib} = [e^{\beta\hbar\omega_0/2} - e^{-\beta\hbar\omega_0/2}]^{-1} ,$$

we are assuming:

(1) The volume is large enough that q_{int} is independent of volume. This is important when distinguishing C_p from C_V, since the difference will be that from q_{trans}, i.e., the ideal gas value.

(2) Excited electronic (or nuclear) states are sufficiently high in energy to be ignored in Q.

(3) Internal nuclear motion is well approximated by a rigid rotor-harmonic oscillator, which neglects:
- anharmonicity in the vibrations
- centrifugal distortion
- rotation-vibrational coupling.

(4) The high temperature formula for q_{rot} assumes that $\Theta_{rot} \ll T$ (i.e., 12.1 K << 300 K) .

From the partition functions, we obtain

$$\langle E\rangle = -\partial \ell n Q/\partial\beta = N\varepsilon_0 + Nk_BT + N\frac{\hbar\omega_0}{2}\coth(\beta\hbar\omega_0/2) + 3/2\, Nk_BT .$$

Therefore,

$$C_V = 5/2\ Nk_B + N\left(\frac{\hbar\omega_0}{2}\right)^2 \operatorname{csch}^2(\beta\hbar\omega_0/2)/\ k_B T^2 ,$$

and $C_p = C_V + Nk_B$

$$= 7/2\ Nk_B + Nk_B\left(\frac{\hbar\omega_0}{2k_B T}\right)^2 \operatorname{csch}^2(\beta\hbar\omega_0/2) .$$

Furthermore,

$$S = \frac{\langle E\rangle - A}{T}$$

$$= Nk_B\Big\{\ 5/2 + \frac{\beta\hbar\omega_0}{2}\coth(\beta\hbar\omega_0/2) + \ln g_{nuc} + \ln(T/\Theta_{rot})$$

$$- \ln[e^{\beta\hbar\omega_0/2} - e^{-\beta\hbar\omega_0/2}] + \ln\Big[\frac{V}{h^3}\Big(\frac{2\pi(m_H + m_{Br})}{\beta}\Big)^{3/2}\Big]$$

$$- \ln N + 1\ \Big\} .$$

The necessary numerical input is:

$T = 298$ K $\qquad \beta\hbar\omega_0 = \frac{3700}{298} \qquad \Theta_{rot} = 12.1$ K $\qquad h = 6.63 \times 10^{-27}$ erg ·s

$$g_{nuc} = \underbrace{(2\,\tfrac{1}{2} + 1)}_{H^1}\ \underbrace{(2\,\tfrac{3}{2} + 1)}_{Br^{81},\ Br^{79}} = 8 \qquad \frac{V}{N} = \frac{k_B T}{p} = 4.06 \times 10^{-2} cm^3$$

yielding

$S = 51.7$ cal/mol K , (Note: CRC value of 47.5 neglects the $\ln g_{nuc}$ contribution of 4.14 cal/mol K)

and

$C_p = 6.97$ cal/mol K .

4.26 $Q(N,V,T) = \frac{q^N}{N!} \qquad q = \frac{V}{h^3}\left(\frac{2\pi m}{\beta}\right)^{3/2}$

(a) $$\Xi = \sum_{N=0}^{\infty} \frac{q^N}{N!} e^{\beta\mu N} = \sum_{N=0}^{\infty} \frac{(qe^{\beta\mu})^N}{N!} = e^{(qe^{\beta\mu})} = e^{zV}$$

$$z = \frac{e^{\beta\mu}}{h^3} \left(\frac{2\pi m}{\beta}\right)^{3/2}$$

(b) $\Xi = e^{\beta pV} \Rightarrow \beta pV = zV \qquad p = z/\beta$

$$\langle N \rangle = \frac{\partial \ell n \Xi}{\partial \beta\mu} = Vz \quad \text{implies} \quad z = \rho$$

Hence, $p = \rho/\beta$, the ideal gas law.

(c) $$\frac{\sqrt{\langle(\delta\rho)^2\rangle}}{\rho} = \frac{\sqrt{\langle(\delta N)^2\rangle}}{\langle N \rangle}$$

$$\langle(\delta N)^2\rangle = \frac{\partial \langle N \rangle}{\partial \beta\mu} = V \frac{\partial z}{\partial \beta\mu} = Vz = \langle N \rangle$$

$$\frac{\sqrt{\langle(\delta\rho)^2\rangle}}{\rho} = \frac{1}{\langle N \rangle^{1/2}} = \sqrt{k_B T/pV} = 2.01 \times 10^{-10} .$$

(d) To second order in δN

$$\ell n P(N) = \ell n P(\langle N \rangle) + \frac{1}{2} (\delta N)^2 \left(\frac{\partial^2 \ell n P(N)}{\partial N^2}\right)_{N = \langle N \rangle} ,$$

i.e., assuming a Gaussian distribution around $\langle N \rangle$ with variance $\langle(\delta N)\rangle^2$, we get

$$\frac{P(N)}{P(\langle N \rangle)} \approx e^{-10^{-12}\langle N \rangle/2} = e^{-10^7},$$ which is <u>very</u> unlikely .

5.1 Since the number of nearest neighbors in a D-dimensional cubic lattice is 2D,

$$E_0 = - JN\left(\frac{1}{2} \cdot 2D\right) = - DNJ ,$$

where the ½ corrects for double counting pairs. Note that for a two-dimensional <u>triangular</u> lattice, each spin has 6 nearest neighbors, so that in that case

$$E_0 = - JN\left(\frac{1}{2} 6\right) = - 3NJ .$$

5.3 $$H = - J \sum_{i=1}^{N} s_i s_{i+1} = - J \sum_{i=1}^{N} b_i$$

where $b_i = s_i s_{i+1} = \pm 1$ is the bond between spin i and i+1. The b_i's completely determine the s_i's up to the value of s_1. This degeneracy is related to the fact that only N-1 of the b_i's are independent:

$$\prod_{i=1}^{N} b_i = \prod_{i=1}^{N} s_i^2 = 1 .$$

However, as $N \to \infty$, the contribution in energy due to the Nth bond $b_N = \prod_{i=1}^{N-1} b_i$ is small compared to the other bonds $\sum_{i=1}^{N-1} b_i$. So

$$Q = \sum_{\{s_i = \pm 1\}} e^{\beta J \sum_{i=1}^{N} s_i s_{i+1}} \underset{N\to\infty}{\approx} \sum_{\{b_i = \pm 1\}} e^{\beta J \sum_{i=1}^{N} b_i}$$

$$= \prod_{j=1}^{N} \left(\sum_{b=-1,1} e^{\beta J b}\right) = \left(e^{\beta J} + e^{-\beta J}\right)^N = [2\cosh(\beta J)]^N .$$

Note: With some patience, the reader can compute Q for finite N without neglecting edge effects. Also of interest is the transfer matrix method illustrated in the solution to Exercise 5.21.

5.4 Note that the Ising model energy can be written as

$$- mH \sum_{i=1}^{N} (2n_i-1) - J \sum_{i,j}' (2n_i-1)(2n_j-1)$$

$$= - 2mH \sum_{i=1}^{N} n_i + mNH - 4J \sum_{i,j}' n_i n_j + \underbrace{2J \sum_{i,j}' (n_i + n_j)}_{\text{same as } z \sum_{i=1}^{N} n_i} - \underbrace{J \sum_{i,j}' 1}_{Nz/2}$$

As a result, the answer is

$$Q_{ISING}(\beta,N,H;J,m) = \exp[\beta N(mH - Jz/2)] \, \Xi_{\substack{LATTICE\\GAS}}(\beta,N,\mu;\varepsilon)$$

with $\mu = 2mH - 2Jz$, $\varepsilon = 4J$ and z is the number of nearest neighbor sites to any given lattice site.

5.6 $$m = \tanh(\beta\mu H + \beta zJm) = \frac{e^{2(\beta\mu H + \beta zJm)} - 1}{e^{2(\beta\mu H + \beta zJm)} + 1}$$

or

$$me^{2(\beta\mu H + \beta zJm)} + m = e^{2(\beta\mu H + \beta zJm)} - 1$$

or

$$2\beta(\mu H + zJm) = \ln\left(\frac{1+m}{1-m}\right) \quad \text{which implies} \quad \beta = \frac{1}{2Jzm} \ln\left(\frac{1+m}{1-m}\right), \text{ at } H = 0 .$$

Now,

$$\ln\left(\frac{1+m}{1-m}\right) = \ln(1+m) - \ln(1-m)$$

$$= \left(m - \frac{m^2}{2} + \frac{m^3}{3}\right) - \left(-m - \frac{m^2}{2} - \frac{m^3}{3}\right) + \ldots$$

$$= 2m + 2\,m^3/3 + \ldots,$$

which is the Taylor expansion of $\ell n\left(\frac{1+m}{1-m}\right)$ about $m_c = 0$.

So $\beta = \frac{1}{2Jzm}\,(2m + 2\,m^3/3 + \ldots)$

or

$$\frac{1}{3}\,m^2 \approx Jz(\beta - \beta_c)$$

Thus, for $T \to T_c$,

$$m \propto (T_c - T)^{1/2} \quad \text{or } \beta = 1/2 .$$

In reading this solution, try not to confuse the reciprocal temperature β with the critical exponent β.

Finally, as $T \to 0$, $\beta \to \infty$, it's easiest to analyze

$$m = \tanh(\beta zJm) = \frac{e^{\beta zJm} - e^{-\beta zJm}}{e^{\beta zJm} + e^{-\beta zJm}}$$

which clearly tends to -1 as $\beta \to \infty$ if $m < 0$, and $+1$ as $\beta \to \infty$ if $m > 0$.

The complete mean field solution is plotted in the figure.

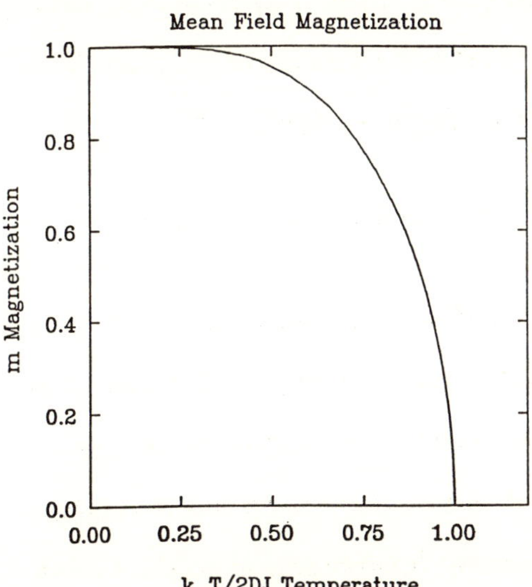

In this figure, only positive m are plotted since the curve is symmetric about the temperature axis. Note that $k_B T/2DJ$ is $T/T_c^{(MF)}$, where $T_c^{(MF)}$ is the mean field theory prediction of T_c.

5.8 In the mean field theory, different spins are uncorrelated. Thus

$$\begin{aligned}\langle E\rangle_{\rm MF} &= -\mu H \sum_i \langle s_i\rangle_{\rm MF} - \frac{1}{2}\sum_{i,j} J_{ij}\langle s_i s_j\rangle_{\rm MF} \\ &= -\mu H \sum_i \langle s_i\rangle_{\rm MF} - \frac{1}{2}\sum_{i,j} J_{ij}\langle s_i\rangle_{\rm MF}\langle s_j\rangle_{\rm MF} ,\end{aligned}$$

The desired result follows from the association $\langle s_i\rangle_{\rm MF} = m$. Alternatively, you may begin from the perspective of Sec.(5.5) where the partition function in this approximation is shown to be

$$\ln Q \approx \ln Q_{\rm MF} - \beta\langle \Delta E\rangle_{\rm MF}$$

where $-\beta\langle\Delta E\rangle_{\rm MF}$ is given by Eq.(d) of p. 138, and

$$Q_{\rm MF} = [2\cosh(\beta\mu H + \beta z J m)]^N .$$

The desired result for the internal energy is obtained from differentiating this approximation for $\ln Q$ with respect to $-\beta$. Here, in the differentiation, note that m is a function of β; in fact, from Eq.(b) on p. 136 with $\Delta H = zJm/\mu$, we find

$$\frac{dm}{d\beta} = (\mu H + zJM)(1-m^2)/[1-\beta z J(1-m^2)]$$

for $T < T_c$. For $T > T_c$ and $H = 0$, mean field theory gives $m = 0$, and thus predicts zero internal energy in that case. This prediction is incorrect since (although the average coupling to the field is zero — i.e., there is no broken symmetry and long ranged order) there are still correlations between nearest neighbor spins.

5.18 $g(K') = 2g(K) - h(K)$ where $h(K)$ is analytic at $K = K_c$,

$K' = K'(K) = 3/8 \ln \cosh(4K)$ $K'(K_c) = K_c$, and $h(K_c) = g(K_c)$.

Assume $C \propto |K - K_c|^{-\alpha}$ near $K = K_c$,

that is, $C = \{$ a less singular part $\} + a|K - K_c|^{-\alpha}$.

The second term with $\alpha > 0$ dominates as $K \to K_c$.

Then since $c = \frac{d^2}{dk^2} g(K)$,

we can write $g(K) = \{$ a less singular part $\} + b|K - K_c|^{2-\alpha}$

where b is a constant. Plugging this into the first equations above, we get

$$\{\text{a less singular part}\} + b|K' - K_c|^{2-\alpha} = 2[\{\text{a less singular part}\} + b|K - K_c|^{2-\alpha}] - h(K) \ .$$

As $K \to K_c$, the dominant contributions are

$$b|K' - K_c|^{2-\alpha} = 2b|K - K_c|^{2-\alpha} \ .$$

Taylor expanding $K'(K)$ about $K = K_c$ gives

$$K' = K_c + \frac{\partial K'}{\partial K}\Big|_{K_c} \cdot (K - K_c) \ .$$

Hence,

$$\left| \frac{\partial K'}{\partial K}\Big|_{K_c} \cdot (K - K_c) \right|^{2-\alpha} = 2|K - K_c|^{2-\alpha} ,$$

or

$$\left| \frac{\partial K'}{\partial K}\Big|_{K_c} \right|^{2-\alpha} = 2 \ , \quad \text{which implies } (2-\alpha) \ln \frac{\partial K'}{\partial K}\Big|_{K_c} = \ln 2 \ ,$$

or

$$\alpha = 2 - \ln 2 / \ln \frac{\partial K'}{\partial K}\Big|_{K_c} \quad .$$

Note that the actual value of h(k) is irrelevant.

5.21 (a) $$\mathrm{Tr}(\underset{\sim}{q}^N) = \sum_{s_1, s_2 \ldots, s_N = \pm 1} \prod_{j=1}^{N} e^{(s_j + s_{j+1})h/2 + s_j s_{j+1} K}$$

$$= \sum_{s_1, s_2, \ldots, s_N = \pm 1} \exp[\sum_{j=1}^{N} (hs_j + Ks_j s_{j+1})] = Q$$

(b) $$\mathrm{Tr}(\underset{\sim}{q}^N) = \mathrm{Tr}[\begin{pmatrix} \lambda_+ & 0 \\ 0 & \lambda_- \end{pmatrix}^N] = \mathrm{Tr}\,[\begin{pmatrix} \lambda_+^N & 0 \\ 0 & \lambda_-^N \end{pmatrix}] = \lambda_+^N + \lambda_-^N$$

(c) $$\underset{\sim}{q} = \begin{pmatrix} a & c \\ c & b \end{pmatrix}$$

$$|\lambda \underset{\sim}{I} - \underset{\sim}{q}| = (\lambda - a)(\lambda - b) - c^2 = \lambda^2 - (a+b)\lambda + ab - c^2 = 0$$

implies

$$\lambda = \frac{a+b}{2} \pm \sqrt{(\frac{a-b}{2})^2 + c^2} \quad .$$

Thus,

$$\lambda_\pm = e^K \cosh(h) \pm \sqrt{e^{2K} \sinh^2(h) + e^{-2K}}$$

$$= e^K [\cosh(h) \pm \sqrt{\sinh^2(h) + e^{-4K}}\,] \; .$$

Further,

$$Q = \lambda_+^N + \lambda_-^N = \lambda_+^N [1 + (\frac{\lambda_-}{\lambda_+})^N] \; , \text{ so}$$

$$\frac{\ln Q}{N} = \ln \lambda_+ + \frac{1}{N} \ln [1 + (\frac{\lambda_-}{\lambda_+})^N] \;\to\; \ln \lambda_+ \; , \quad \text{as } N \to \infty \quad .$$

Hence,

$$\frac{\ln Q}{N} = \ln(\lambda_+) = K + \ln[\cosh(h) + \sqrt{\sinh^2(h) + e^{-4K}}] \; .$$

(d) $$\langle s_1\rangle = \frac{\partial}{\partial h}\left(\frac{\ln Q}{N}\right) = \frac{\sinh(h) + \dfrac{\sinh(h)\cosh(h)}{\sqrt{\sinh^2(h) + e^{-4K}}}}{\cosh(h) + \sqrt{\sinh^2(h) + e^{-4K}}}$$

$$\lim_{h\to 0}\langle s_1\rangle = \lim_{h\to 0} \sinh(h)\cdot\left(\frac{1 + \cosh(h)/\sqrt{\sinh^2(h) + e^{-4K}}}{\cosh(h) + \sqrt{\sinh^2(h) + e^{-4K}}}\right)$$

$$= 0 \quad .$$

5.22 (a)

$Q(K,h,N) =$

$$\sum_{\{\text{odd } s_i\}} \left(e^{(h/2)[s_1+s_3]+h+K[s_1+s_3]} + e^{(h/2)[s_1+s_3]-[h+K(s_1+s_3)]}\right) \times \ldots.$$

$$\ldots \times \left(e^{(h/2)[s_{N-2}+ s_N]+h+K[s_{N-2}+s_N]} + e^{(h/2)[s_{N-2}+s_N]-[h+K(s_{N-2}+s_N)]}\right)$$

$$= [f(K,h)]^{N/2}\, Q(K',h',N/2) \quad .$$

Thus,

$$e^{(h/2)(s_1+s_3)+h+K(s_1+s_3)} + e^{(h/2)(s_1+s_3) - [h+K(s_1+s_3)]}$$

$$= f(K,h)[e^{(h'/2)(s_1+s_3) + K's_1s_3}]$$

For all the possible states of s_1 and s_3:

$$s_1 = s_3 = 1 \ : \ f(K,h)\, e^{h'+ K'} = e^{2(h+K)} + e^{-2K} \qquad (1)$$

$$s_1 = -s_3 \ : \ f(K,h)e^{-K'} = e^{h} + e^{-h} \qquad (2)$$

$$s_1 = s_3 = -1 \ : \ f(K,h)e^{-h'+ K'} = e^{-2K} + e^{-2h + 2K} \ . \qquad (3)$$

By multiplying Eqs (1),(3) with Eq(2) twice quickly yields

$$f(K,h) = 2[\cosh^2(h)\cosh(2K+h)\cosh(-2K+h)]^{1/4} \ .$$

Division of Eq(1) by Eq(3) quickly gives

$$h' = h + \frac{1}{2} \ell n \left[\frac{\cosh(2K+h)}{\cosh(-2K+h)} \right] .$$

Division of Eq(1) times Eq(3) by Eq(2) twice easily leads to

$$K' = \ell n[\cosh(2K+h)\cosh(-2K+h)/\cosh^2(h)] / 4 .$$

(b) For $K, h > 0$: $|2K + h| > |-2K + h|$, so that $\frac{\cosh(2K+h)}{\cosh(-2K+h)} > 1$,

implying $h' > h$. Further,

$$e^{4K'} = \frac{\cosh(2K + h)\cosh(-2K + h)}{\cosh^2(h)}$$

$$= e^{4K} \frac{(e^h + e^{-4K-h})(e^{-4K+h} + e^{-h})}{(e^h + e^{-h})(e^h + e^{-h})} < e^{4K}$$

Hence, $K' < K$.

So, the flows as written cause h to increase and K to decrease as illustrated in the picture.

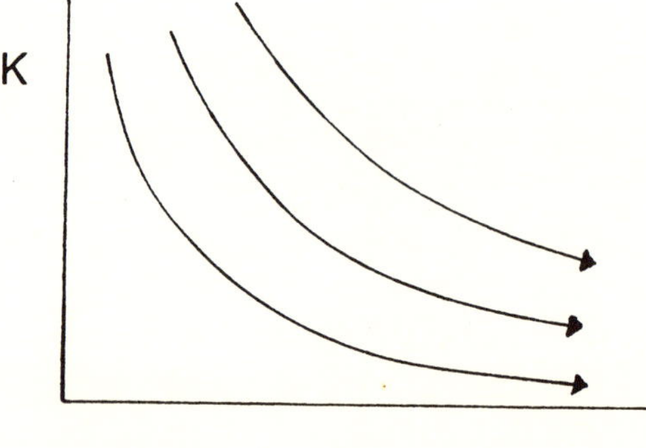

(c) $g(h,K) = \frac{1}{N} \ln Q$

$$g(h,K{=}0) = \frac{1}{N} \ln(e^{h} + e^{-h})^{N}$$

$$= \ln(2\cosh(h)) \quad .$$

As written in part (a) and (b), flow from $g(h,K = 0.01)$ will be towards the h-axis, away from $(h,K) = (1,1)$. We would like then to invert the parameter flow, i.e., find $h = h(h',K')$; $K = K(h',K')$.

Using these inverses, our errors will diminish with each iteration, since then $g(h,K) = 1/2\, g(h',K') + 1/2\, f(h',K')$.

One method to get reversed flows (from low K to high K) is to first flow forward from (1,1) (to lower K values), and then retrace the steps taken back up to (1,1):

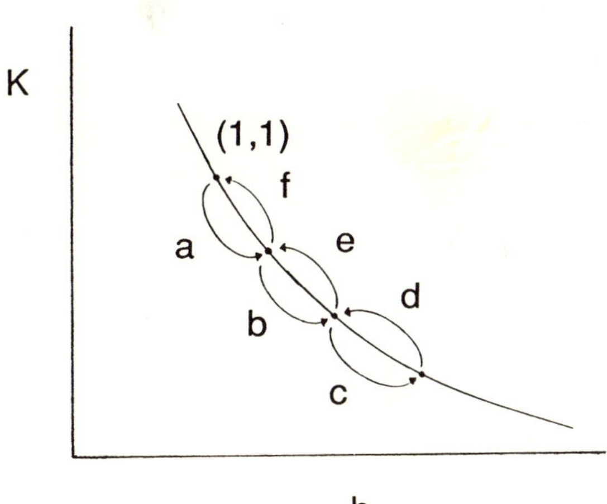

On the reverse trip, d-e-f, we can use our low K "accurate" approximation for g, the error halving with each iteration.

g	h	K
2.00285	1.00	1.00
2.41	1.94	0.469
2.84	2.81	2.23×10^{-2}
2.86	2.86	7.09×10^{-6}
2.86	2.86	7.5×10^{-12}

Generate starting at (h,K) = (1,1) and moving down according to flows.

Start at bottom with

$$g(h = 2.86,\ K = 7.5 \times 10^{-12}) \approx g(2.86,\ 0) = \ln[2\cosh(2.86)]\ ,$$

and iterate upwards using

$$g(h,K) = \frac{1}{2}\, g(h',\ K') + \frac{1}{2}\ln f(h',K')\quad .$$

The convergence to h = 2.86, K = 0 exhibits a stable fixed point, and in fact any $h = h_0$, K = 0 is a fixed point in flow.

The exact answer is (from Exercise 5.22)

$$\frac{\ln Q(K=1,\ h=1)}{N} = K + \ln\left(\cosh(h) + \sqrt{\sinh^2(h) + e^{-4K}}\ \right) = 2.00285$$

which matches the RG value of 2.00285 above.

An analytic inverse in closed form can be found by using e^{2h} and e^{4k} as variables, which reduces the problem to the finding of the roots of a 4th degree polynomial.

5.24 Parts (a) and (b): The rules for construction of legal configurations admit only three energy levels:

(i) $E = \infty$ if any partition has other than two atoms. Now, given that there are two atoms per partition, we can have

(ii) $\cdots \begin{smallmatrix}0\\0\end{smallmatrix}\Big| \;\begin{smallmatrix}0\\0\end{smallmatrix}\Big| \;\begin{smallmatrix}0\\0\end{smallmatrix}\Big| \cdots$ (or $\cdots \Big|\begin{smallmatrix}0\\0\end{smallmatrix} \;\Big|\begin{smallmatrix}0\\0\end{smallmatrix} \;\Big|\begin{smallmatrix}0\\0\end{smallmatrix} \;\Big|\begin{smallmatrix}0\\0\end{smallmatrix} \cdots$) .

Having two atoms on the same side of a partition means that the two atoms in the next cell must belong to the next partition. The degeneracy is 2, and the energy is $E = N \cdot 0 = 0$.

(iii) The only other noninfinite energy states are those where each partition has an atom on each side:

$^0\big|_0$ means that the next partition is either $^0\big|_0$ or $_0\big|^0$,

and the choice between these two possibilities for each partition is unrelated to other partitions, e.g.,

$\cdots {}^0\big|_0 \;\; {}^0\big|_0 \;\; {}_0\big|^0 \;\; {}^0\big|_0 \cdots$

These states have energy $E = N\varepsilon$, and the degeneracy is 2^N.

(c) As a result of the preceding analysis,

$$Q = 2 + 2^N e^{-\beta N\varepsilon} .$$

(d) $$\frac{A}{N} = \frac{\ell nQ}{-\beta n} = \frac{\ell n\left(2 + (2e^{-\beta\varepsilon})^N\right)}{-\beta N}$$

Therefore,

$$\lim_{N\to\infty} \frac{A}{N} = \begin{cases} -k_BT\,\ell n2 + \varepsilon, & 2e^{-\beta\varepsilon} > 1 , \\ 0 , & 2e^{-\beta\varepsilon} < 1 . \end{cases}$$

Hence

$$\frac{\langle E\rangle}{\langle N\rangle} = \frac{1}{\langle N\rangle}\frac{\partial \ln Q}{\partial(-\beta)} = \begin{cases} \varepsilon, & 2e^{-\beta\varepsilon} > 1 \\ 0, & 2e^{-\beta\varepsilon} < 1 . \end{cases}$$

(e) A phase transformation is associated with $\beta = \frac{1}{\varepsilon} \ln 2$, or

$$T_0 = \frac{\varepsilon}{k_B \ln 2} .$$

This problem illustrates how phase transitions can arise due to competition between entropic and energetic effects.

5.26 (a) $$Q = \lim_{P\to\infty} (\varepsilon\Delta)^{P/2} \sum_{\{u_i\}} \exp[\sum_{i=1}^{P} \kappa\, u_i u_{i+1}] \int_{-\infty}^{\infty} d\mathbf{E} \exp[-\beta \mathbf{E}^2/2\sigma + (\varepsilon\mu \sum_{i=1}^{P} u_i)\mathbf{E}] .$$

The integral can be evaluated to be

$$\exp\{\beta/2\sigma[\tfrac{\sigma}{\beta} (\varepsilon\mu \sum_{i=1}^{P} u_i)]^2\} \int_{-\infty}^{\infty} d\mathbf{E}\, e^{-\beta/2\sigma[\mathbf{E}-\sigma/\beta(\varepsilon\mu \sum_{i=1}^{P} u_i)]^2}$$

$$= \exp[\frac{\beta\sigma\mu^2}{2P^2} (\sum_{i=1}^{P} u_i)(\sum_{j=1}^{P} u_j)] \cdot \sqrt{\frac{2\sigma}{\beta}} \cdot \sqrt{\pi} .$$

So,

$$Q = \sqrt{\frac{2\pi\sigma}{\beta}} \lim_{P\to\infty} \{(\varepsilon\Delta)^{P/2} \sum_{\{u_i\}} e^{\sum_{i=1}^{P} \kappa\, u_i u_{i+1} + (\frac{\beta\mu^2\sigma}{2P^2}) \sum_{i,j=1}^{P} u_i u_j}\} .$$

(b) From the text, we know

$$Q(\mathbf{E}) = \lim_{P\to\infty} \sum_{\{u_i\}} (\varepsilon\Delta)^{P/2} e^{\sum_{i=1}^{P} [\kappa\, u_i u_{i+1} + h u_i]}$$

$$= 2 \cosh[\beta\sqrt{\Delta^2 + \mu^2\mathbf{E}^2}] .$$

Hence,

$$\langle m \rangle = \mu \langle u_1 \rangle = \int_{-\infty}^{\infty} d\mathbf{E}\, e^{-\beta \mathbf{E}^2/2\sigma} \frac{\partial Q(\mathbf{E})}{\partial(\beta \mathbf{E})} / Q$$

$$= \frac{\mu \int_{-\infty}^{\infty} d\mathbf{E}\, e^{-\beta \mathbf{E}^2/2\sigma}\, 2\sinh\left(\beta\sqrt{\Delta^2 + \mu^2\mathbf{E}^2}\right) \cdot \mu\mathbf{E}/\sqrt{\Delta^2 + \mu^2\mathbf{E}^2}}{\int_{-\infty}^{\infty} d\mathbf{E}\, e^{-\beta \mathbf{E}^2/2\sigma}\, 2\cosh\left(\sqrt{\Delta^2 + \mu^2\mathbf{E}^2}\right)} .$$

As $\beta \to \infty$, $\cosh\beta x, \sinh\beta x \to e^{\beta x}$, so

$$\langle m \rangle = \frac{\mu \int_{-\infty}^{\infty} d\mathbf{E}\, \exp[-\beta\mathbf{E}^2/2\sigma + \beta\sqrt{\Delta^2 + \mu^2\mathbf{E}^2}]\left(\mu\mathbf{E}/\sqrt{\Delta^2 + \mu^2\mathbf{E}^2}\right)}{\int_{-\infty}^{\infty} d\mathbf{E}\, \exp[-\beta\mathbf{E}^2/2\sigma + \beta\sqrt{\Delta^2 + \mu^2\mathbf{E}^2}]}$$

$$\equiv \frac{\mu \int_{-\infty}^{\infty} d\mathbf{E}\, e^{-\beta W(\mathbf{E})}\left(\mu\mathbf{E}/\sqrt{\Delta^2 + \mu^2\mathbf{E}^2}\right)}{\int_{-\infty}^{\infty} d\mathbf{E}\, e^{-\beta W(\mathbf{E})}} ,$$

where the last equality defines $W(\mathbf{E})$ and the weighting function $e^{-\beta W(\mathbf{E})}$. As $\beta \to \infty$, the weighting function becomes a delta function $\propto \delta(\mathbf{E} - \mathbf{E}_{min})$ where $\mathbf{E}_{min}$ is the $\mathbf{E}$ that minimizes $W(\mathbf{E})$. Exploiting this idea is a method of "steepest descent." Note

$$\frac{\partial W}{\partial \mathbf{E}} = \frac{\mathbf{E}}{\sigma} - \frac{\mu^2\mathbf{E}}{\sqrt{\Delta^2 + \mu^2\mathbf{E}^2}} \quad \text{which implies } \mathbf{E}_{min} = 0 \text{ or } \sigma\mu^2 = \sqrt{\Delta^2 + \mu^2\mathbf{E}_{min}^2} .$$

In the latter case,

$$\mathbf{E}_{min}^2 = \sigma^2\mu^2 - \Delta^2/\mu^2 \quad \text{which has the real solutions}$$

$$\mathbf{E}_{\pm} = \pm\, \mu\left(\sigma - \frac{\Delta}{\mu^2}\right)^{1/2}$$

when $\sigma\mu^2/\Delta > 1$. When this condition is met, we have broken symmetry. In other words, $\mathbf{E}_{\pm}$ starts contributing to $W(\mathbf{E})$ at $\sigma\mu^2/\Delta = 1$. For $\sigma\mu^2/\Delta < 1$, $\langle m \rangle = 0$, while for $\sigma\mu^2/\Delta > 1$, as

$\beta\to\infty$, $\langle m\rangle \neq 0$. This emergence of broken symmetry will cause $\langle(\delta m)^2\rangle$ to diverge at $\sigma = \sigma_{crit} = \Delta/\mu^2$. More is said about this broken symmetry in the isomorphic problem, Exercise 5.27.

5.27 (a) Doing the sum over spins with the transfer matrix method (Exercise 5.21):

$$\sum_{\{s_i=\pm 1\}} \exp\left[\sum_{i=1}^{N} \beta h s_i + \beta J s_i s_{i+1}\right] = \mathrm{Tr}(\underset{\sim}{q}^N) \text{ , where } \underset{\sim}{q} = \begin{pmatrix} e^{\beta h+\beta J} & e^{-\beta J} \\ e^{-\beta J} & e^{-\beta h+\beta J} \end{pmatrix}$$

Now, the eigenvalues of $M = \begin{pmatrix} a & c \\ c & b \end{pmatrix}$ are $\lambda = \left(\frac{a+b}{2}\right) \pm \sqrt{\left(\frac{a-b}{2}\right)^2 + c^2}$,

$$\text{so } \mathrm{Tr}(q^N) = \mathrm{Tr}\begin{pmatrix} \lambda_+^N & 0 \\ 0 & \lambda_-^N \end{pmatrix} = \lambda_+^N + \lambda_-^N \text{ ,}$$

where $\lambda_+ > \lambda_-$ are the eigenvalues of $\underset{\sim}{q}$, and

$$Q(h;\beta,N) = e^{-\beta N h^2/2\sigma} \left(\lambda_+^N + \lambda_-^N\right) \text{ .}$$

Thus,

$$\tilde{A}(h;\beta,N) = -1/\beta \ \ln Q(h;\beta,N)$$

$$= -1/\beta \left[N \ln\lambda_+ + \ln\left(1 + \left(\frac{\lambda_-}{\lambda_+}\right)^N\right) - \beta N h^2/2\sigma\right] \text{ .}$$

For large N, therefore,

$$\tilde{A}(h;\beta,N) = -\frac{N}{\beta} \ln\lambda_+ + \frac{Nh^2}{2\sigma}$$

$$= N\left[\frac{h^2}{2\sigma} - J - \frac{1}{\beta} \ln\left(\cosh(\beta h) + \sqrt{\sinh^2(\beta h) + e^{-4\beta J}}\right)\right] \text{ .}$$

(b) At this point, notice that $\tilde{A}(h;\beta,N)$ is <u>even</u> in h , i.e., if h minimizes $\tilde{A}$, so does -h . Further,

$$\lim_{|h|\to\infty} \tilde{A} \to N\left[\frac{h^2}{2\sigma} - J - |h|\right] \text{ .}$$

Thus, we're looking for bistability in $\tilde{A}$ like this:

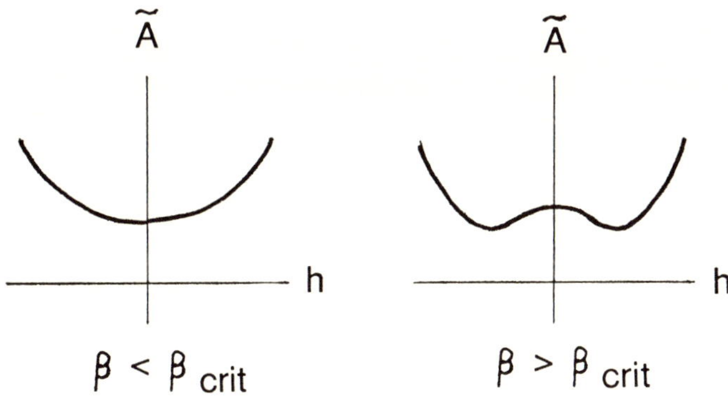

Since $\tilde{A}$ is even and $\tilde{A} \underset{h\to\infty}{\to} \infty$, $\tilde{A}$ will have a bistability in h when $\left.\frac{\partial \tilde{A}}{\partial h}\right|_{h'} < 0$ for $h' > 0$ near zero; i.e., $\tilde{A}$ is concave downward at h = 0 when $\partial\tilde{A}/\partial h' < 0$ for h' near zero. Hence, $\partial\tilde{A}/\partial h = 0$ can be used to determine the critical temperature. Alternatively, note that taking the $\beta\to\infty$ limit gives

$$\lim_{\beta\to\infty} \cosh(\beta h) = e^{\beta|h|}, \quad \lim_{\beta\to\infty} \sinh(\beta h) = \text{sign}(h)\cdot e^{\beta|h|}, \quad \text{and hence}$$

$$\tilde{A}(h, \beta\to\infty, N) = N\left[\frac{h^2}{2\sigma} - J - \frac{1}{\beta}\ln(e^{\beta|h|})\right]$$

$$= N\left[\frac{h^2}{2\sigma} - J - |h|\right]$$

and does show bistability. However, we don't get an equation for β_{crit} from this analysis.

(c) Near h = 0 for h > 0, we find

$$\left.\frac{\partial\tilde{A}}{\partial h}\right|_{h=0^+} = N\left[\frac{h}{\sigma} - \sinh(\beta h)\, e^{2\beta J}\right]$$

$$= Nh\left[\frac{1}{\sigma} - \beta e^{2\beta J}\right], \qquad \text{since } \sinh(\beta h) \approx \beta h .$$

This derivative changes in sign as a function of β when

$$\beta_{crit}\, e^{2\beta_{crit}J} = \frac{1}{\sigma}$$

which is an invertible function in β since it's montonically increasing.

Another way to view the analysis of this Exercise is to solve for $\frac{\partial\tilde{A}}{\partial h} = 0$ to find the stable minima in $\tilde{A}(h)$. This equation can be written $\tilde{A}'(h_{min},\beta) = 0$, an implicit function for $\beta = \beta(h_{min})$. β_{crit} is then the limiting value of $\beta = \beta(h_{min})$, as $h_{min} \to 0$, and so we look for $\left.\frac{\partial\tilde{A}}{\partial h}\right|_{h\to 0^+} = 0$ and solve for β_{crit}.

Chapter 6. Monte Carlo Methods in Statistical Mechanics

To perform Exercises 6.10, 6.11 and 6.14-6.16, you will collect statistics from long Monte Carlo runs. These exercises help illustrate what must be done to obtain reliable information from such calculations. In this regard, efficient and fast computer codes are important, and as emphasized in the text, much can be done to improve the programs listed in the text. At the very least, one should use a compiled version of the codes. As a standard of excellence, however, you might want to view Jeffrey Fox's Monte Carlo of the Ising magnet.* It utilizes binary arithmetic only and machine language, and it runs from 10^2 to 10^3 times faster on a PC than the Ising model program listed in the text. As well as providing rapid access to good statistics, the greater speed permits one to view rather large systems in real time. The views are both instructive and pleasing to the eye.

Some new books on Monte Carlo methods in particular and computer simulations in general are worthy additions to the bibliography of this chapter:

M. H. Kalos and P. A. Whitlock, Monte Carlo Methods Vol. I: Basics, (Wiley Interscience, New York, 1986);

H. Gould and J. Tobochnik, An Introduction to Computer Simulation Methods, Parts 1 and 2, (Addison-Wesley, Reading, 1988);

M. P. Allen and D. J. Tildesley, Computer Simulation of Liquids, (Oxford U. Press, Oxford, 1987).

The last of these is especially pertinent to the material covered in Chapters 7 and 8.

---

*J. Fox, "Fox's Ising Model Simulation for PC Compatibles," placed in the public domain on 3/25/88.

6.10 Above T_c, we find c_{ij} decays more or less exponentially as i gets farther from j, and random thermal fluctuations destroy long-ranged correlations. In fact $c_{ij} \to 0$ for i far from j until $T \to T_c$ from above, at which point $c_{ij} \neq 0$.

Below T_c, we have broken symmetry, and as $T \to 0$, we find $|\langle s_1 \rangle|$ is significantly greater than zero. This is an indication that we are frozen into either the up state or the down state. One result of this broken symmetry is that as our runs apparently show $\langle s_1 \rangle^2 \to 1$, and hence $\langle s_i s_j \rangle \to 1$ as well, we will observe $c_{ij} \to 0$ as $T \to 0$.

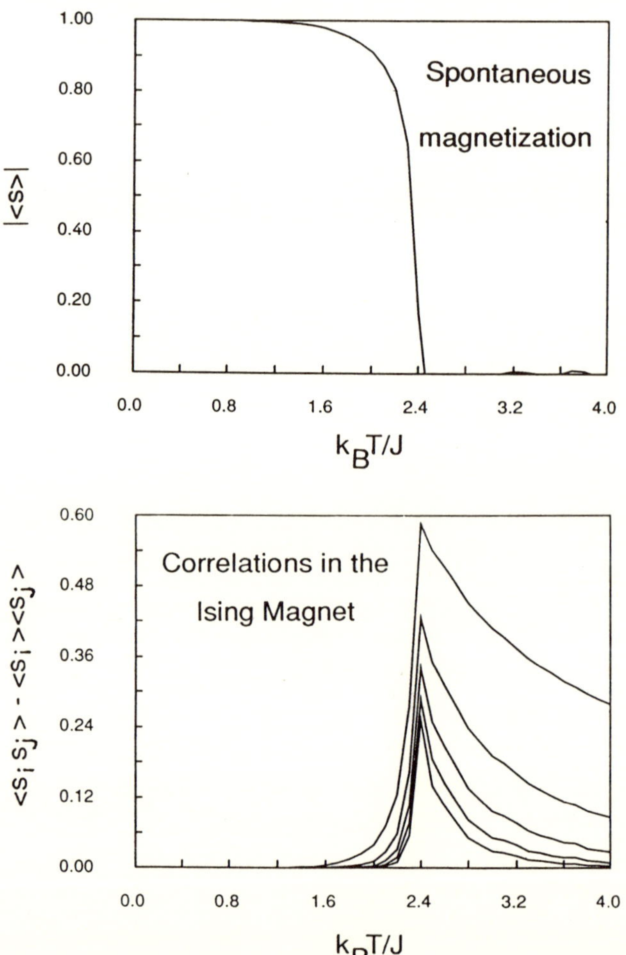

Fig. Representative results for exercise 6.10. The lattice was equilibrated for 100 passes, and then statistics were compiled for 5,000 passes. Spin-spin correlations were measured up to the fifth nearest neighbor.

6.11 The effect of the interface is to introduce long-ranged correlations between spins along the interface, even while $T \ll T_c$ and the correlations in the bulk are only short-ranged. However, as defined in the code given in the book, the ground state of the 20 × 20 lattice, even with the biasing field on, has all the spins aligned the same way (with the exception of the biased spins). Simply setting the initial configuration to be an interface would sample a metastable well. To observe the interface fluctuations we can increase the bulk of the lattice relative to the interface by changing the dimensions to be 40 × 20 .

Note that in implementing the calculation, there are four columns of spins each that can be considered bulk or interface. In our calculations the dependence of $\langle s \rangle$ and $\langle s_i s_j \rangle$ on the distance from the biasing field is averaged over. Correlations shown in the graph are $\langle s_i s_j \rangle - \langle s_i \rangle \langle s_j \rangle$.

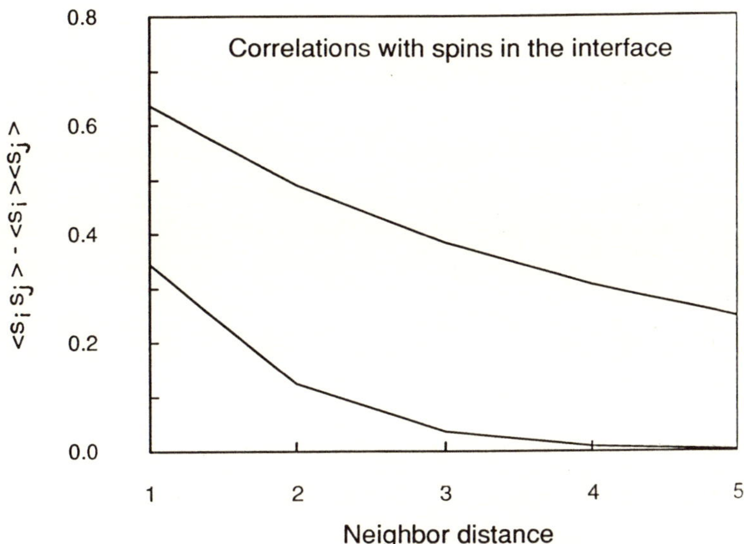

Fig. These representative results were obtained on a 40 × 20 lattice at $k_BT/J = 1.2$ using 100 passes to equilibrate and 10,000 passes to measure correlations. The biasing field keeps spins 1-20 up and 21-40 down in the top row. Spins in columns 1, 20, 21 and 40 are considered to be in the interface, and columns 10, 11, 30 and 31 were taken for bulk measurements. The upper line shows correlations along the interface, and the lower line shows the correlation of a spin at the interface into the bulk. Correlations between bulk spins are ~ 10^{-3} or 10^{-4} and are not discernible on this graph.

6.12 A straightforward algorithm is to choose α large enough, or equivalently, to scale the range of the uniform random numbers [0,1] to $[-x_{max}, x_{max}]$ so that most of the Gaussian will fit in this range. Then trial x's will be accepted with a probability $\propto p(x)$, normalized so that for all x, $prob(x) \leq 1$. The trouble is that the fraction of accepted trial x's tends to be small. This fraction is just that of the area under the curve to the rectangle:

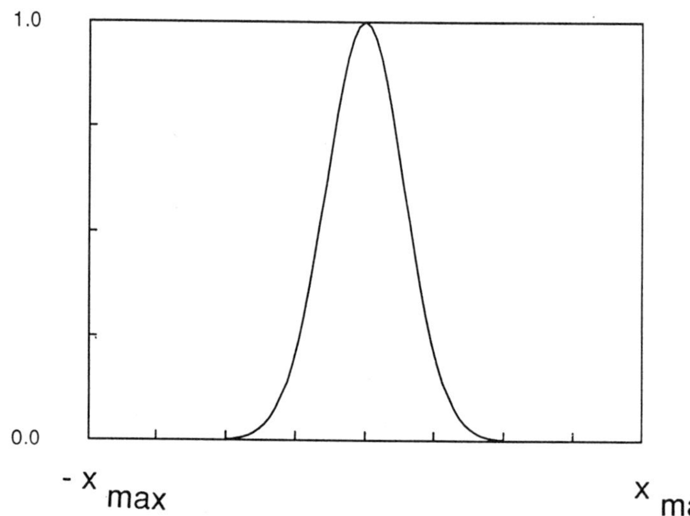

In other words, if we pick x uniformly from the interval, we will then reject most choices. This inefficient method does not exploit any form of importance sampling. There are a couple of possibilities for improvement.

One is to imagine that the distribution $p(x) = \sqrt{\alpha/\pi}\ e^{-\alpha x^2}$ is due to an energy E(x), where $\beta E(x) = \alpha x^2$, and to employ the usual Metropolis algorithm with diffusive motion. Then trial samples will tend to come from the regions of lower E(x) and higher probability. The drawback is that one random number x_{i+1} is correlated with the previous one x_i from which it made its move to $x_{i+1} = x_i + \delta x$, with δx small. Nonetheless, the calculated moments should be correct.

Another method, equivalent to the "force bias" Monte Carlo of problem 6.13, is to change $\pi_{\nu\nu'}$ to reflect the more likely regions of p(x), thereby increasing the $A_{\nu\nu'}$'s .

The first algorithm described above corresponds to $\pi_{xx'}$ = constant and $A_{xx'} \propto e^{-\alpha x'^2}$. As we said, that scheme is not very efficient. If instead we make a change of variables, such as $x = f(y)$, with y picked uniformly from [0,1] then the probability of picking x in $[x_0, x_0 + dx]$ is

$$\left|\frac{dy}{dx}\right|_{x=x_0} \cdot dx \propto \left|\frac{df^{-1}}{dx}\right|_{x=x_0} ,$$

where $y = f^{-1}(x)$. So if we want to achieve 100% acceptance

$A_{xx'} = 1$, then our trials must be chosen with $\pi_{xx'} = \sqrt{\alpha/\pi}\ e^{-\alpha x^2}$

and thus

$$\left.\frac{df^{-1}}{dx}\right|_{x_0} = \sqrt{\alpha/\pi}\ e^{-\alpha x^2} \text{ which implies } f^{-1}(x_0) = \sqrt{\alpha/\pi}\int_0^{x_0} e^{-\alpha x^2} dx = \mathrm{erf}(x_0) ,$$

i.e., f is the inverse of the error function. Note this scheme only generates positive x; the sign should be assigned randomly.

In some problems, such a perfect solution is not possible. But as long as $A_{xx'}$ increases and begins to approach unity, we have made good progress. In the present context, any change of variables favoring sampling smaller x is better. For example, let

$x = \frac{1-y}{cy}$ where $y \in [0,1]$ is uniformly picked. This implies

$y = \frac{1}{1 + cx}$.

So the x are picked with probability

$$\pi_x \propto \frac{c}{(1+cx)^2}$$, and so we must accept with probability

$$A_x \propto \frac{(1+cx)^2}{c} \sqrt{\pi/\alpha}\ e^{-\alpha x^2}$$

to preserve w_x . The proportionalities are chosen so that $\pi_x \leq 1$ and $A_x \leq 1$. Again only positive x are generated and the signs should be randomly chosen. Note only one index is retained because any given choice is completely independent of previous choices.

6.13 (a) The condition of detailed balance, $w_{\nu\nu'}/w_{\nu'\nu} = \exp(-\beta\Delta E_{\nu\nu'})$, with $\Delta E_{\nu\nu'} = E_{\nu'} - E_{\nu}$, gives

$$\frac{\pi_{\nu\nu'}A_{\nu\nu'}}{\pi_{\nu'\nu}A_{\nu'\nu}} = \exp(-\beta\Delta E_{\nu\nu'}), \quad \text{or} \quad A_{\nu\nu'} = \frac{\pi_{\nu'\nu}A_{\nu'\nu}}{\pi_{\nu\nu'}} \exp(-\beta\Delta E_{\nu\nu'}) .$$

Since $A_{\nu\nu'}$ is a probability, $A_{\nu\nu'} \leq 1$.

So if $\frac{\pi_{\nu'\nu}}{\pi_{\nu\nu'}} e^{-\beta E_{\nu\nu'}} < 1$, we let $A_{\nu\nu'} = \frac{\pi_{\nu'\nu}}{\pi_{\nu\nu'}} e^{-\beta\Delta E_{\nu\nu'}}$ and $A_{\nu'\nu} = 1$; otherwise, we can switch indices, $\nu \leftrightarrow \nu'$, i.e.,

$$A_{\nu'\nu} = \frac{\pi_{\nu\nu'}}{\pi_{\nu'\nu}} e^{-\beta\Delta E_{\nu'\nu}} \text{ and } A_{\nu\nu'} = 1 .$$

(b) To make a transition every move, we

let $A_{\nu\nu'} = A_{\nu'\nu} = 1$. So then

$$w_{\nu\nu'} = \pi_{\nu\nu'} = \begin{cases} p, & \nu' = \nu+1 \\ 1-p, & \nu' = \nu-1 \end{cases} .$$

This form, which is independent of ν, is possible because of the symmetry of the problem — each pair of adjacent levels is equivalent to any other pair. With detailed balance, designating $E_{\nu'} > E_{\nu}$, $\nu' = \nu+1$, we therefore have,

$$\frac{w_{\nu\nu'}}{w_{\nu'\nu}} = e^{-\beta\Delta E_{\nu\nu'}} = e^{-\beta\hbar\omega} = \frac{p}{1-p} \equiv x$$

or

$$p = \frac{1}{1 + x^{-1}} = \frac{1}{1 + e^{\beta\hbar\omega}} .$$

Note: in the limits $\hbar\omega \to 0$, $p \to 1/2$,

and $\hbar\omega \to \infty$, $p \to 0$.

6.14 (a) $$Q = \int_{-\infty}^{\infty} d\mathbf{E}[\text{Tr } e^{-\beta(H_0 - m(\mathbf{E} + \mathbf{E}_{app}))}]\, e^{-\beta\mathbf{E}^2/2\sigma}$$

$$\langle m\rangle = \frac{\partial \ln Q}{\partial(\beta\mathbf{E}_{app})} = \frac{1}{Q}\frac{\partial Q}{\partial(\beta\mathbf{E}_{app})} = \frac{\int_{-\infty}^{\infty} d\mathbf{E} e^{-\beta\mathbf{E}^2/2\sigma}\text{Tr } m\, e^{-\beta H}}{Q} .$$

Evaluating the trace gives

$$Q = \int_{-\infty}^{\infty} d\mathbf{E}\, e^{-\beta\mathbf{E}^2/2\sigma}\, 2\cosh[\beta(\Delta^2 + \mu^2(\mathbf{E} + \mathbf{E}_{app})^2)^{1/2}]$$

$$= 2(\Delta/\mu)\int_{-\infty}^{\infty} d\overline{\mathbf{E}}\, e^{-\overline{\beta\mathbf{E}}^2/2L} \cosh(\overline{\beta}\xi) \text{ , and so}$$

$$\frac{\partial Q}{\partial(\beta\mathbf{E}_{app})} = 2\int_{-\infty}^{\infty} d\mathbf{E}\, e^{-\beta\mathbf{E}^2/2\sigma} \sinh(\overline{\beta}\mathbf{E})\ \frac{\overline{\beta}}{\beta}\ \frac{1}{2}\ \xi^{-1}\ 2(\overline{\mathbf{E}} + \overline{\mathbf{E}}_{app})(\mu/\Delta)$$

$$= 2(\Delta/\mu)\int_{-\infty}^{\infty} d\overline{\mathbf{E}}\, e^{-\overline{\beta\mathbf{E}}^2/2L} \sinh(\overline{\beta}\xi)\ \mu\, (\overline{\mathbf{E}} + \overline{\mathbf{E}}_{app})/\xi$$

from which the desired formula follows directly.

6.16 The modification to include windows is simple. Since we are basically adding a potential that is infinite outside the window (i.e., not under the umbrella) and zero otherwise, we need only check if we've stepped outside the window. This is done at the time we also check $e^{-\beta\Delta E} > rnd$. Moves outside the window are rejected. We can scan through each window for the minimum in A(M). Once one is found for a given T, we can assume for T' near T , that M'_{min} is near M_{min}, so we need only scan a few neighboring windows.

However, as $T \to T_c$, it will be hard to find the minimum since a wide range of M are equally probable. We find that the barrier drops faster than the minimum moves in as illustrated in the schematic figure.

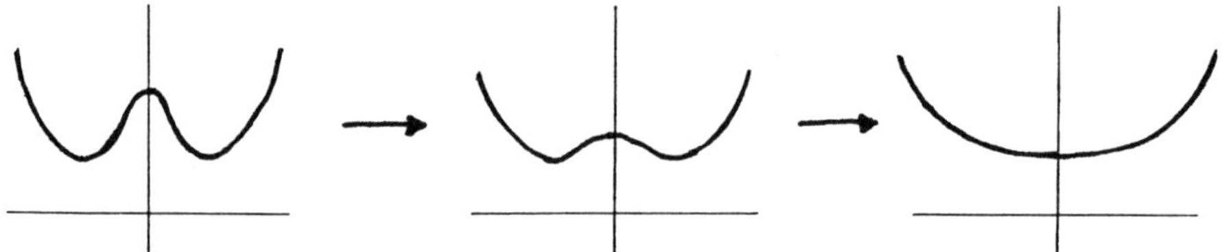

Notice that the criteria of a vanishing bistability can be used unambiguously to define a critical temperature for finite systems. Of course, such a definition leads to system size dependent T_c which becomes the true critical temperature only as $N = L \times L \to \infty$. So, our results for the "phase diagram," $(\frac{M_{min}}{L^2})$ vs T is shown in the figure for several values of L. The $L \to \infty$ behavior of M_{min}/L^2 is the exact M(T), and we've plotted M(T) on the same graph for comparison.

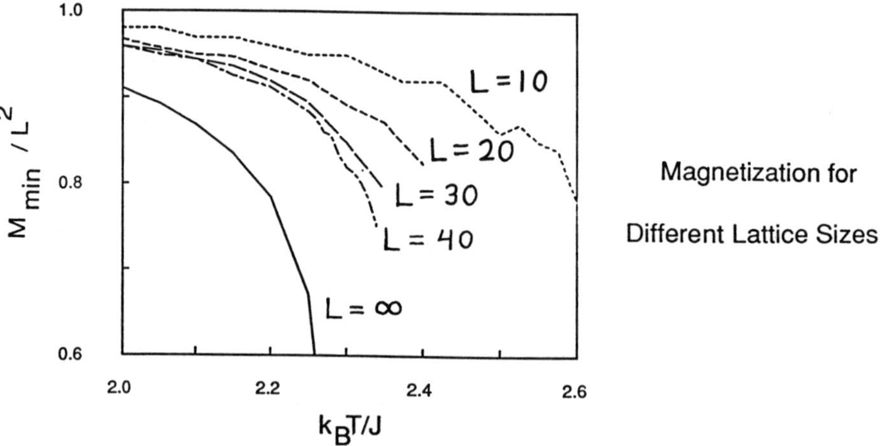

Fig. These results were obtained with 5000 passes/window, and each window had the width $L^2/20$ except for the L = 40 case where we used $L^2/40$ window widths.

As described above, we determined T_c for each value of L by extrapolating to $M_{min} = 0$. The results of those extrapolations are given in the table. We then extrapolated T_c to the $L \to \infty$ by using a procedure known as <u>finite size scaling</u>. In particular, we assumed

$$[T_c - T_c(L)]^b \propto 1/L$$

where $T_c \equiv T_c(\infty)$. We determined the exponent b from the slope of $[T_c - T_c(L)]$ vs. ℓn L . This procedure gave $b \approx 1.0 \pm 0$. Then a plot of $T_c(L)$ vs. 1/L gave

$$T_c \approx 2.26 \pm 0.02$$

as the $1/L \to 0$ intercept. This estimate can be compared with Onsager's $T_c = 2.269$.

L	$T_c(L)$
10	2.60
20	2.43
30	2.38
40	2.34

A discussion of the finite size scaling method is given in Sec. 1.2 of Binder and Stauffer's Chapter in <u>Applications of the Monte Carlo Method in Statistical Physics</u>, ed. K. Binder (Springer-Verlag, N.Y., 1984). Also see Part 2 of Gould and Tobochnik's <u>Computer Simulation Methods</u> referred to at the beginning of this manual chapter.

7.10 The kinetic energy and pair potential parts of <E> will be the same as discussed in the text. For the three-body term, we write

$$\left\langle \sum_{i>j>\ell=1}^{N} u^{(3)}(\mathbf{r}_i - \mathbf{r}_j, \mathbf{r}_j - \mathbf{r}_\ell) \right\rangle = \frac{1}{6} N(N-1)(N-2) \langle u^{(3)}(\mathbf{r}_{12},\mathbf{r}_{23})\rangle$$

$$= \frac{N(N-1)(N-2)}{6} \frac{\int dr^N u^{(3)}(\mathbf{r}_{12},\mathbf{r}_{23})\, e^{-\beta U(r^N)}}{\int dr^N\, e^{-\beta U(r^N)}}$$

$$= \frac{1}{6} \frac{\int d\mathbf{r}_1 \int d\mathbf{r}_2 \int d\mathbf{r}_3\, u^{(3)}(\mathbf{r}_{12}, \mathbf{r}_{23})\, N(N-1)(N-2) \int dr^{N-3} e^{-\beta U(r^N)}}{\int dr\, e^{-\beta U(r^N)}}$$

$$= \frac{1}{6} \int d\mathbf{r}_1 \int d\mathbf{r}_2 \int d\mathbf{r}_3\, \rho^{(3/N)} (\mathbf{r}_1, \mathbf{r}_2, \mathbf{r}_3)\, u^{(3)}(\mathbf{r}_{12}, \mathbf{r}_{23}) .$$

For a uniform (homogeneous) system, $\rho^{(3/N)}(\mathbf{r}_1, \mathbf{r}_2, \mathbf{r}_3) = \rho^3 g^{(3)}(\mathbf{r}_{12}, \mathbf{r}_{23})$, and we can integrate out one of the degrees of freedom to get the volume:

$$= \frac{1}{6} V \int d\mathbf{r}_{12} \int d\mathbf{r}_{23}\, \rho^3 g^{(3)}(\mathbf{r}_{12}, \mathbf{r}_{23})\, u^{(3)}(\mathbf{r}_{12}, \mathbf{r}_{23})$$

$$= N \cdot \frac{1}{6} \rho^2 \int d\mathbf{r}_{12} \int d\mathbf{r}_{23}\, g^{(3)}(\mathbf{r}_{12}, \mathbf{r}_{23})\, u^{(3)}(\mathbf{r}_{12}, \mathbf{r}_{23}) .$$

Thus

$$\frac{\langle E\rangle}{N} = \frac{3}{2} k_B T + \frac{1}{2} \rho \int d\mathbf{r}\; g(r)u(r)$$

$$+ \frac{1}{6} \rho^2 \int d\mathbf{r}_{12} \int d\mathbf{r}_{23}\, g^{(3)}(\mathbf{r}_{12}, \mathbf{r}_{23})\, u^{(3)}(\mathbf{r}_{12}, \mathbf{r}_{23}) .$$

7.11 $$\beta p = \left(\frac{\partial \ln Q}{\partial V}\right)_{T,N} = \frac{\partial}{\partial V} \ln \int_V dr^N\, e^{-\beta U(r^N)}$$

We change coordinates to $\mathbf{x}_i = V^{-1/3}\, \mathbf{r}_i$, thereby accounting for each of the three cartesian components, $d\mathbf{x}_i = V d\mathbf{r}_i$. Hence,

$$\beta p = \frac{\partial}{\partial V} \ln V^N \int_1 dx^N \exp[-\beta \sum_{i>j=1}^{N} u(V^{1/3}\, x_{ij})] \quad ,$$

where the integration is now over the unit volume, not the system volume. Differentiation is now easy, directly affecting the pair potential. Accounting for the sum over N(N-1)/2 equivalent pairs, differentiation yields

$$\beta p = \rho - \frac{\beta}{6V} \frac{\int dx^N N(N-1)\exp[-\beta \sum_{i>j=1}^{N} u(V^{1/3}x_{ij})][du(V^{1/3}x_{12})/d(V^{1/3}x_{12})]\, x_{12}V^{1/3}}{\int dx^N \exp[-\beta \sum_{i>j=1}^{N} u(V^{1/3}x_{ij})]}$$

$$= \rho - \frac{\beta}{6V} \int dr^N N(N-1)\exp[-\beta \sum_{i>j=1}^{N} u(r_{ij})][du(r_{12})/dr_{12}]r_{12} \Big/ \int dr^N \exp[-\beta \sum_{i>j=1}^{N} u(r_{ij})]$$

$$= \rho - \frac{\beta}{6V} \int d\underset{\sim}{r}_1 \int d\underset{\sim}{r}_2 \frac{du(r_{12})}{dr_{12}}\, r_{12}\, N(N-1) \int dr^{N-2}\, e^{-\beta U(r^N)} \Big/ \int dr^N e^{-\beta U(r^N)}$$

$$= \rho - \frac{\beta}{6V} \int d\underset{\sim}{r}_1 \int d\underset{\sim}{r}_2 \frac{du(r_{12})}{dr_{12}}\, r_{12}\, \rho^{(2/N)}(\underset{\sim}{r}_1, \underset{\sim}{r}_2) \quad .$$

Therefore, for a uniform system

$$\frac{\beta p}{\rho} = 1 - (\beta\rho/6) \int d\underset{\sim}{r}\; g(r) r\; du(r)/dr \quad . \qquad \text{Q.E.D.}$$

7.12 $\int d\underset{\sim}{r}[e^{-\beta u(r)} - 1] = 4\pi \int_0^\infty dr\; r^2[e^{-\beta u(r)} - 1]$

$$= \frac{4\pi r^3}{3} [e^{-\beta u(r)} - 1] \Big|_0^\infty - 4\pi \int_0^\infty \frac{r^3}{3}\, e^{-\beta u(r)} (-\beta \frac{du}{dr})\, dr$$

As $r \to \infty$, $u(r) \to 0$, and

$$\frac{r^3}{3} [e^{-\beta u(r)} - 1] \to \frac{r^3}{3} [-\beta u(r)] \to 0$$

for quickly enough decaying u(r), i.e., faster than r^{-3}. Therefore, the boundary term in the integration by parts gives zero. As a result,

$$-\frac{1}{2}\int d\underset{\sim}{r}[e^{-\beta u(r)}-1] = \frac{1}{2}\left(-\frac{\beta}{3}\right)\int_0^\infty dr\ 4\pi\ r^2\ e^{-\beta u(r)}\ r\ \frac{du}{dr}$$

$$= -\ \beta/6\int d\underset{\sim}{r}\ e^{-\beta u(r)}\ r\ \frac{du}{dr}\ . \qquad \text{Q.E.D.}$$

7.13 For hard spheres

$$B_2(T) = -\ 2\pi[-\int_0^\sigma dr\ r^2] = \frac{2\pi\sigma^3}{3} \equiv b_0\ .$$

For a square well

$$B_2(T) = -\ 2\pi\left\{-\int_0^\sigma dr\ r^2 + \int_\sigma^{\sigma'} dr\ r^2[e^{\beta\varepsilon}-1]\right\}$$

$$= \frac{2\pi}{3}[\sigma'^3 - e^{\beta\varepsilon}(\sigma'^3-\sigma^3)]\ .$$

At the Boyle temperature,

$$\frac{2\pi}{3}[\sigma'^3 - e^{\beta_B\varepsilon}(\sigma'^3-\sigma^3)] = 0\ ,$$

therefore,

$$\varepsilon/k_BT_B = \ell n\ \frac{\sigma'^3}{\sigma'^3-\sigma^3}$$

or

$$T_B = \frac{\varepsilon}{k_B\ell n\ (\sigma'^3/(\sigma'^3-\sigma^3))}\ .$$

7.14 The figure shows a graph of the square well B_2 with σ' adjusted such that

$$\int_\sigma^\infty dr\ r^2 u_{LJ}(r) = -\ \varepsilon\int_\sigma^{\sigma'} dr\ r^2$$

which implies $\sigma' = 9^{1/3}\sigma$. With this choice of σ', the square well B_2

should be a reasonable estimate of $B_2^{(LJ)}$. However, unlike the square well B_2, $B_2^{(LJ)}$ vanishes at very large temperatures since $e^{-\beta u(r)} - 1 \to 0$ as $\beta \to 0$.

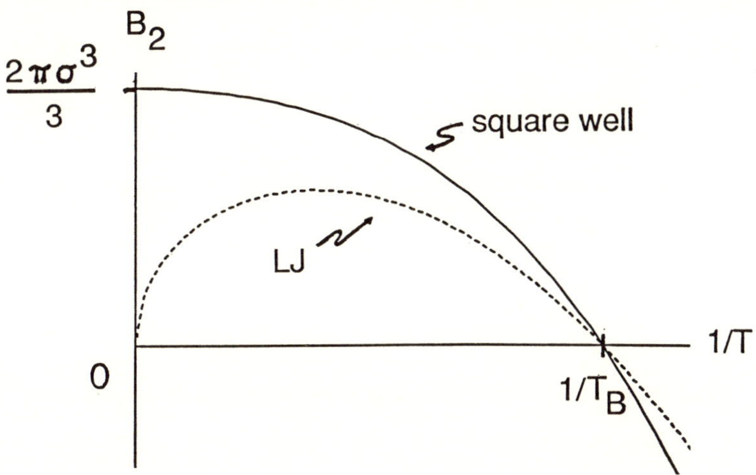

7.28 The velocity or momentum averages are isotropic and factorable from the coordinates. Therefore, only one average is necessary to answer all parts to this Exercise.

(a) $$\langle v_x^2 \rangle = \int d\underset{\sim}{p}\, v_x^2 \; {}^{-\beta p^2/2m} \Big/ \int d\underset{\sim}{p}\; e^{-\beta p^2/2m}$$

$$= \frac{\int_{-\infty}^{\infty} dp_x (p_x/m)^2 \, e^{-\beta p_x^2/2m}}{\int_{-\infty}^{\infty} dp_x \, e^{-\beta p_x^2/2m}}$$

since the p_x degree of freedom is statistically uncorrelated from the others. Performing the integral gives

$$\langle v_x^2 \rangle = \frac{1}{\beta m} \quad .$$

Alternatively,

$$\langle v_x^2 \rangle = \frac{2}{m} \, \frac{1}{2} \, m \, \langle v_x^2 \rangle = \frac{2}{m} \, \frac{1}{3} \, \langle p^2/2m \rangle$$

$$= (2/3m) \, \frac{3}{2} \, k_B T = 1/\beta m \quad .$$

(b) $\langle v_x^2\, v_y^2\rangle = \langle v_x^2\rangle\langle v_y^2\rangle = (\frac{1}{\beta m})^2$

(c) $\langle v^2\rangle = \langle v_x^2 + v_y^2 + v_z^2\rangle = 3\langle v_x^2\rangle = \frac{3}{\beta m}$

(d) $\langle v_x\rangle = 0$

(e) $\langle(v_x + bv_y)^2\rangle = \langle v_x^2\rangle + 2b\langle v_x\rangle\langle v_y\rangle + b^2\langle v_y^2\rangle$

$$= (1+b^2)\ \langle v_x^2\rangle$$

$$= \frac{1+b^2}{\beta m}$$

Classically, these averages won't be affected by isothermal volume/pressure changes (i.e. through the potential) since the momentum distribution is dependent only on the temperature. When the classical assumption breaks down at extremely high pressures, one would then need to consider the quantum dispersion of particles since the dispersion (i.e., the uncertainty principle) leads to statistical correlations between positions and momenta.

7.29 (a) Since the Hamiltonian is separable, one can factor the partition function as discussed in Chapter 4. This gives

$$Q = \left(\frac{e^{-\beta\hbar\omega/2}}{1 - e^{-\beta\hbar\omega}}\right)^{3N} = \left(e^{\beta\hbar\omega/2} - e^{-\beta\hbar\omega/2}\right)^{-3N}$$

where we have used the fact that $(\frac{1}{2} + n)\,\hbar\omega$ with $\omega = \sqrt{k/m}$ is the nth energy level for a one-dimensional harmonic oscillator.

(b) In the classical world, the partition function still factors,

$$Q = \frac{1}{h^{3N}} \int dr^N \int dp^N\, e^{-\beta H(r^N,\, p^N)}$$

$$= \frac{1}{h^{3N}} [(\int_{-\infty}^{\infty} dx\, e^{-\beta k(x - x_0)^2/2}) \cdot (\int_{-\infty}^{\infty} dp_x\, e^{-\beta p_x^2/2m})]^{3N}$$

$$= \frac{1}{h^{3N}} [\sqrt{\frac{2\pi}{\beta K}} \cdot \sqrt{\frac{2m\pi}{\beta}}]^{3N}$$

$$= (\beta\hbar\omega)^{-3N} .$$

(c) Since $e^{\pm\beta\hbar\omega/2} \sim 1 \pm \beta\hbar\omega/2$ for small β ,

$$Q_{QM} \sim [(1 + \frac{\beta\hbar\omega}{2}) - (1 - \frac{\beta\hbar\omega}{2})]^{-3N}$$

$$= (\beta\hbar\omega)^{-3N} = Q_{classical} .$$

7.31 (a) For $\rho \to \rho_{CP}$ (but $\rho < \rho_{Cp}$) we approach the g(x) for a close packed system, i.e., delta functions spaced ℓ apart. The lower the density, ρ, the broader the peaks, and the more rapid the decay to unity at large x.

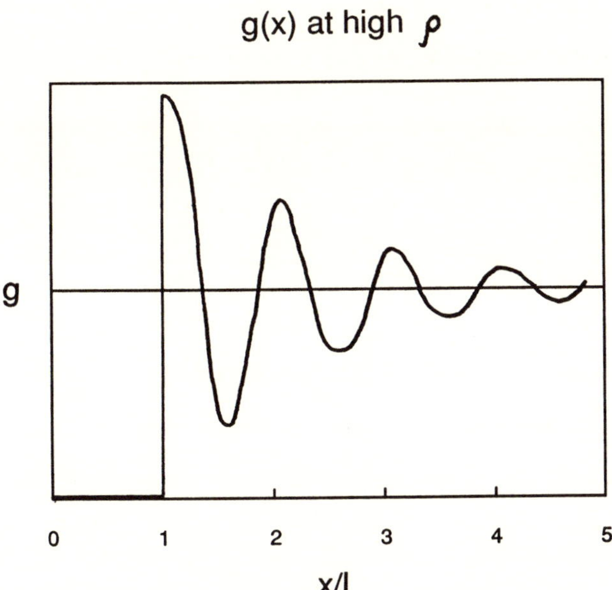

(b) Since $u(x)$ is ∞ for $|x| < \ell$ and zero otherwise, $g(x)$ is therefore a step function at low ρ. As $\rho \to 0$, $g(x) = e^{-\beta u(x)}$.

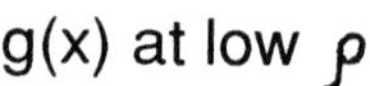

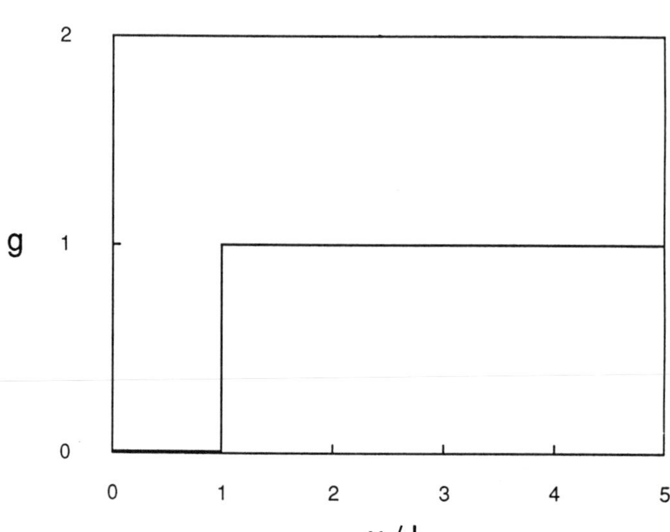

(c) Notice for a hard potential system, $\beta u(x)$ is either infinite or zero, so $\beta u(x)$ is independent of temperature. Thus <u>g(x) is independent of temperature.</u> The configurational properties of hard core potential systems have no temperature.

(d) Since the integral of $\rho g(x)$ determines the number of neighbors, and since there will be one neighbor between 0 and $(3/2)\ell$, we have

$$\int_0^{(3/2)\ell} dx\, g_{CP}(x) = \rho_{CP}^{-1}$$

$$= \ell \quad .$$

(e) Note that the momenta are uncoupled from the positions, and that the momentum distribution is proportional to $\exp(-\beta p^2/2m)$. Thus,

(i) $\langle v \rangle = 0$.

Further,

(ii) $\langle |v| \rangle = 2 \int_0^\infty \frac{p}{m} \; e^{-\beta p^2/2m} \, dp \Big/ \int_{-\infty}^\infty e^{-\beta p^2/2m} \, dp$

$$= \sqrt{\frac{2}{\pi m \beta}}$$

and

(iii) $\langle \frac{1}{2} mv^2 \rangle = \frac{1}{2\beta} = \frac{1}{2} k_B T$.

Finally,

(iv) $\langle E \rangle = \langle KE \rangle + \langle PE \rangle$

$$= N \cdot \langle \frac{p^2}{2m} \rangle = \frac{1}{2} N k_B T$$

where we have noted that $\langle PE \rangle = 0$ since the "potential energy" is zero for all acceptable configurations.

(f) Since $\beta p = \rho + b\rho^2 + \ldots$, we can identify b as the second virial coefficient. Hence

$$b = -\frac{1}{2} \int_{-\infty}^{\infty} dx[e^{-\beta u(x)} - 1] = -\frac{1}{2} \int_{-\ell}^{\ell} dx(-1) = \ell \ .$$

Note that b is independent of temperature.

(g) The partition function for the system can be calculated exactly. The easiest way is to think of the excluded volume due to the rods ($N\ell$) as being subtracted out of the possible position space of each rod (L). Secondly, configurations which are identical except for the switching of rods are counted as the same state, i.e., we need to divide by a factor of N! . So,

$$Q = \frac{1}{\lambda^N} \frac{1}{N!} \int_0^{L-N\ell} dx_1 \int_0^{L-N\ell} dx_2 \ldots \int_0^{L-N\ell} dx_N$$

$$= \frac{1}{\lambda^N} \frac{1}{N!} (L - N\ell)^N \ .$$

This value for Q can also be arrived at from scaling and changes of coordinates of the configuration integral: First, adopting one particular ordering of the rods,

$$Q = \frac{1}{\lambda^N} \int_0^L dx_1 \int_0^L dx_2 \cdots \int_0^L dx_N \, e^{-\beta U}$$

$$= \frac{1}{\lambda^N} \int_{\ell}^{L-(N-1)\ell} dx_1 \int_{x_1+\ell}^{L-(N-2)\ell} dx_2 \cdots \int_{x_{N-1}+\ell}^{L} dx_N \, ,$$

which becomes after a change in variables to $x_i' = x_i - i\ell$,

$$Q = \frac{1}{\lambda^N} \int_0^{L-N\ell} dx_1' \int_{x_1'}^{L-N\ell} dx_2' \cdots \int_{x_{N-1}'}^{L-N\ell} dx_N'$$

$$= \frac{1}{\lambda^N} \frac{1}{N!} (L - N\ell)^N \, .$$

Then

$$\beta p = \left.\frac{\partial \ell n Q}{\partial L}\right|_{T,N} = N \frac{1}{L-N\ell} = \frac{\rho}{1-\rho\ell} \, .$$

Thus, $b = \ell$, as already noted.

7.32 (a) Neglecting the internal structure in the low density limit, the gaseous Ar is an ideal structureless gas. Thus, for M atoms

$$\beta\mu^{(id)} = -\frac{\partial \ell n Q^{(id)}}{\partial M} = \ell n M - \ell n \left(\frac{V}{h^3}\left(\frac{2\pi m}{\beta}\right)^{3/2}\right)$$

$$= \ell n \left(\frac{\beta h^2}{2\pi m}\right)^{3/2} + \ell n \rho$$

$$\equiv f(\beta) + \ell n \rho \, ,$$

with $\rho = M/V$.

(b) Let $U_W(R^N)$ stand for the total potential energy for all N water molecules with R^N denoting all the coordinates necessary to describe the configurations of all N water molecules. In the low density limit, different argon molecules do not interact with each other, but they do interact with the water molecules. Therefore, quite generally, we can view the total potential energy of the solution as

$$U_w(R^N) + \sum_{i=1}^{M} U(\underset{\sim}{r}_i; R^N)$$

where $\underset{\sim}{r}_i$ is the position of the ith argon atom and there are M such atoms. The partition function is then

$$Q = \exp[-Nf_w(\beta)-Mf(\beta)] \frac{1}{N!} \frac{1}{M!} \int dR^N \int d\underset{\sim}{r}_1 \ldots d\underset{\sim}{r}_M \exp\{-\beta[U_w(R^N) + \sum_{i=1}^{M} U(\underset{\sim}{r}_i;R^N)]\}$$

$$= Q_w \frac{e^{-MF(\beta)}}{M!} \int dR^N e^{-\beta U_w(R^N)} [\int d\underset{\sim}{r}_1 \ldots \int d\underset{\sim}{r}_M \prod_{i=1}^{M} e^{-\beta U(\underset{\sim}{r}_i;R^N)}]/\int dR^N e^{-\beta U_w(R^N)}$$

$$\equiv Q_w \frac{e^{-Mf(\beta)}}{M!} \int d\underset{\sim}{r}_1 \ldots \int d\underset{\sim}{r}_M \langle \prod_{i=1}^{M} e^{-\beta U(\underset{\sim}{r}_i;\, R^N)} \rangle_w$$

where Q_w and $\langle \ldots \rangle_w$ denote the partition function and ensemble average for the <u>pure</u> liquid water solvent. Now, since the argon is dilute, different argon atoms are uncorrelated. Thus, the average of the product can be replaced with the product of the average. Hence, due to indistinguishability,

$$Q = Q_w \frac{e^{-Mf(\beta)}}{M!} V^M \langle e^{-\beta U(\underset{\sim}{r}_1;R^N)} \rangle_w^M$$

$$\equiv Q_w V^M \frac{e^{-M[f(\beta) + \beta\Delta\mu]}}{M!}, \quad \text{where} \quad e^{\beta\Delta\mu} = \langle e^{-\beta U(\underset{\sim}{r}_1,\, R^N)} \rangle_w .$$

Note that since the solvent is isotropic, the last average is independent of $\underset{\sim}{r}_i$.

The desired formula for μ is obtained on differentiation of ℓnQ with respect to M with Q given by the last equation.

Due to $\langle 1 \rangle_w = 1$, it is obvious that $\Delta\mu = 0$ when $U(\underset{\sim}{r}_1;\, R^N)$ is zero.

(c) At phase equilibrium, the chemical potentials are equal. Hence,

$$f(\beta) + \ln\rho_G = f(\beta) + \ln\rho_L + \beta\Delta\mu \, , \quad \text{implying } \rho_G = \rho_L e^{\beta\Delta\mu}$$

and since for an ideal gas, $\beta p = \rho_G$,

$$\beta p = \rho_L e^{\beta\Delta\mu} \, . \qquad \text{Q.E.D.}$$

(d) We can simply plug into the result for $\Delta\mu$ derived in Part b. Alternatively, we can follow the suggestion noting

$\Delta\mu$ = change in Helmholtz free energy due to introducing one argon atom into pure liquid water.

$$= - k_B T \ln[Q(1 \text{ argon, N waters})/Q_W]$$

from which the desired result for $\exp(-\beta\Delta\mu)$ follows immediately. Next, define $\Delta\mu_\lambda$ according to

$$e^{-\beta\Delta\mu_\lambda} = \langle \prod_i e^{-\beta\lambda u_{AW}(|\underset{\sim}{r} - \underset{\sim}{r}_i|)} \rangle_W \, .$$

Then

$$\frac{d\beta\Delta\mu_\lambda}{d\lambda} = \frac{\langle \beta N u_{AW}(|\underset{\sim}{r}-\underset{\sim}{r}_1|) \prod_i e^{-\beta\lambda u_{AW}(|\underset{\sim}{r}-\underset{\sim}{r}_i|)} \rangle_W}{\langle \prod_i e^{-\beta\lambda u_{AW}(|r-r_i|)} \rangle_W}$$

$$= \beta \frac{\int dR^N \, N \, u_{AW}(|\underset{\sim}{r}-\underset{\sim}{r}_1|) \, e^{-\beta U_\lambda}}{\int dR^N \, e^{-\beta U_\lambda}} \equiv \beta\langle N \, u_{AW}\rangle_\lambda$$

where

$$U_\lambda \equiv U_W(R^N) + \sum_{i=1}^{N} \lambda \, u_{AW}(|\underset{\sim}{r}-\underset{\sim}{r}_i|)$$

and $\langle\ldots\rangle_\lambda$ indicates the ensemble average weighted by that λ-potential energy. Thus, from the definition of radial distribution functions,

$$\frac{d\Delta\mu_\lambda}{d\lambda} = \rho_W \int d\underset{\sim}{r}\; g_{AW}(r;\lambda)\; u_{AW}(r) \; .$$

Integration from $\lambda = 0$ to $\lambda = 1$ gives the final desired result.

7.33 To prove the virial theorem in two dimensions, use the same proof as given for Exercise 7.11, except that now, $\underset{\sim}{x}_i = V^{-1/2}\, \underset{\sim}{r}_i$. On differentiation, this brings down a factor of 1/2 rather than 1/3. One therefore finds

$$\beta p/\rho = 1 - (\beta\rho/4) \int d\underset{\sim}{r}\; g(r) r \frac{du(r)}{dr}$$

$$= 1 - (\beta\rho\pi/2) \int_0^\infty dr\, r^2\, g(r) \frac{du(r)}{dr}$$

$$= 1 + (\rho\pi/2) \int_0^\infty dr\, r^2 [y(r) \frac{d}{dr}\, e^{-\beta u(r)}]$$

where we have introduced $y(r)$ defined so that $g(r) = \exp[-\beta u(r)]y(r)$. For hard disks,

$$e^{-\beta u(r)} = 0 \;, \quad r < \sigma$$
$$= 1 \;, \quad r \geq \sigma \; .$$

Thus,

$$\frac{d}{dr}\; e^{-\beta u(r)} = \delta(r-\sigma) \quad .$$

As a result,

$$\beta p/\rho = 1 + \rho\pi/2 \int_0^\infty dr\, r^2 y(r)\, \delta(r-\sigma)$$

$$= 1 + (\rho\pi\sigma^2/2)\; y(\sigma) \quad .$$

Recall that $y(r)$ is the distribution function for cavities. It should be continuous at $r = \sigma$. However for $r > \sigma$, $g(r) = y(r)$ so $y(\sigma) = \lim_{r\to\sigma^+} g(r)$. Hence,

$$\beta p/\rho = 1 + (\rho\sigma^2\pi/2)\; g(\sigma^+) \quad .$$

7.35 (a) Let $U_\lambda = \sum_{i>j} [u_0(r_{ij}) + \lambda u_1(r_{ij})]$. Then from

$$dA/d\lambda = -\frac{1}{\beta}\frac{d\ell nQ}{d\lambda} = -\frac{1}{\beta}\frac{d}{d\lambda}\ell n \int dr^N e^{-\beta U_\lambda}$$

one finds using arguments similar to those in Exercise 7.32 that

$$dA/d\lambda = (\rho^2 V/2) \int d\underset{\sim}{r}\; g_\lambda(r)u_1(r) \;.$$

Integration over λ from 0 to 1 yields the desired result.

(b) $$Q/Q_0 = \int dr^N e^{-\beta(U_0+U_1)} / \int dr^N e^{-\beta U_0}$$

$$= \langle e^{-\beta U_1} \rangle_0 \geq e^{-\beta\langle U_1\rangle_0} \;.$$

Further,

$$\langle U_1\rangle_0 = \langle \sum_{i>j} u_1(r_{ij})\rangle_0 = \langle \frac{N(N-1)}{2} u_1(r_{12}) \rangle_0$$

$$= (\rho^2 V/2) \int d\underset{\sim}{r}\; g_0(r)u_1(r) \;.$$

Therefore, since $A - A_0 = -k_BT\,\ell n(Q/Q_0)$,

$$\frac{A-A_0}{N} \leq \frac{1}{2}\rho \int d\underset{\sim}{r}\; g_0(r)u_1(r) \;.$$

8.7 Since $H_B[z] = \begin{cases} 1 & z > q^* \\ 0 & z < q^* \end{cases}$

and $\delta[q(0) - q^*]$ requires $q(0) = q^*$,

we interpret

$$k_{BA}(0) = k_{BA}(0^+) = \lim_{\varepsilon \to 0^+} k_{BA}(\varepsilon) \quad .$$

In other words, we're looking for the flux through q^* for small enough time that trajectories which are started at q^* essentially do not recross. Note the existence of $v(0)$ guarantees an ε small enough that $q(\varepsilon) = q(0) + \varepsilon v(0) + O(\varepsilon^2)$. Then

$$k_{BA}(0) = x_A^{-1} \langle v(0)\ \delta[q(0) - q^*]\ H_B[q(0) + \varepsilon v(0)] \rangle_{\varepsilon \to 0^+}$$

$$= x_A^{-1} \langle v(0)\ \delta[q(0) - q^*]\ H_B[q^* + \varepsilon v(0)] \rangle_{\varepsilon \to 0^+}$$

$$= x_A^{-1} \langle v(0)\ \delta[q(0) - q^*]\ \Theta[v(0)] \rangle$$

where Θ is the Heaviside step function. We can then write

$$k_{BA}(0) = x_A^{-1} \frac{\langle v(0)\ \Theta[v(0)]\ \delta[q(0)-q^*] \rangle}{\langle \delta[q(0) - q^*] \rangle} \langle \delta[q(0) - q^*] \rangle \ .$$

The first term is a conditional probability given $q(0) = q^*$. However, since $q(0)$ is statistically independent of $v(0)$ in the ensemble, we can drop the condition. Note that the original expression can't be factored in this way since $v(t) = v[t;q(0),v(0)]$.

So then

$$k_{BA}(0) = x_A^{-1} \langle v(0)\Theta[v(0)] \rangle \langle \delta[q(0) - q^*] \rangle \ ,$$

and since the distribution for $v(0)$ is even and the time origin is arbitrary,

$$k_{BA}(0) = \langle|v|\rangle \langle\delta(q-q^*)\rangle/2x_A \; .$$

Recognizing $H_B^{(TST)}[q(t)] = \Theta[v(0)]$ gives the transition state theory expression for $k_B^{(TST)}$.

8.13 $$\frac{d}{dt}\,\Delta R^2(t) = 2\int_0^t dt'\langle \underset{\sim}{v}(0)\cdot\underset{\sim}{v}(t')\rangle = 2\int_0^t dt'\langle v^2\rangle\, e^{-t'/\tau}$$

$$= 2\langle v^2\rangle\, \tau[1 - e^{-t/\tau}] \; .$$

Integration and enforcing the initial condition of $\Delta R^2(0) = 0$ gives

$$\Delta R^2(t) = 2\langle v^2\rangle\tau t + 2\langle v^2\rangle\tau^2[e^{-t/\tau} - 1] \; .$$

For $t \ll \tau$, to 2nd order in t,

$$\Delta R^2(t) = 2\langle v^2\rangle\tau t + 2\langle v^2\rangle\tau^2[1 - \frac{t}{\tau} + \frac{t^2}{2\tau^2} - 1]$$

$$= \langle v^2\rangle t^2 \approx \langle(v_0 t)^2\rangle \quad ,$$

i.e., inertial (non-diffusive) behavior. For $t \gg \tau$,

$$\Delta R^2(t) \to 2\langle v^2\rangle\tau t - 2\langle v^2\rangle\tau^2 = 2\langle v^2\rangle\tau(t-\tau)$$

which represents the long time diffusive behavior. The function is plotted for all time in the figure.

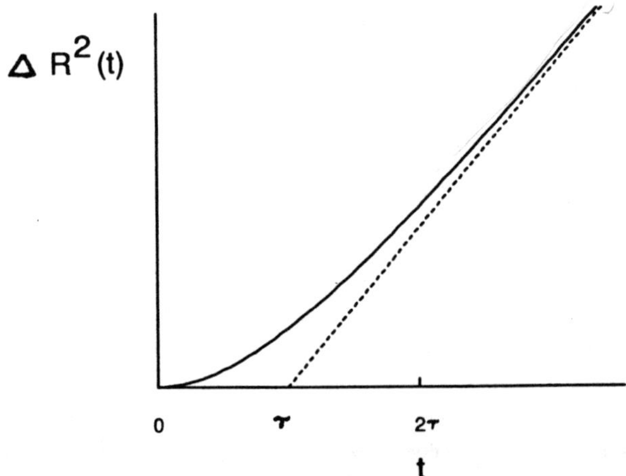

Since

$$D = \frac{1}{3} \int_0^\infty dt \, \langle \underset{\sim}{v}(0) \cdot \underset{\sim}{v}(t) \rangle = \frac{1}{6} \frac{d}{dt} \Delta R^2(t) \Big|_{t\to\infty} = \frac{1}{3} \langle v^2 \rangle \tau$$

then

$$\tau = \frac{3D \, \frac{1}{2} \, m}{\langle \frac{1}{2} \, mv^2 \rangle} = \beta m D \ .$$

For Ar, MW = 40 g/mole, so at T = 80K

$$\tau \approx 60 \text{ fsec} \ .$$

8.20 The primary variable x(t) drives the bath, and the driven bath affects the primary variable through the force

$$f(t) = f_b(t) + \int_{-\infty}^{\infty} dt' \ \chi_b(t-t') \ x(t')$$

$$= f_b(t) - \beta \int_{-\infty}^{t} dt' \ [\frac{d}{d(t-t')}] \ C_b(t-t')] \ x(t')$$

$$= f_b(t) + \beta C_b(0) \ x(t) - \beta C_b(t) \ x(0)$$

$$- \beta \int_0^t dt' \ C_b(t-t') \ \dot{x}(t') \qquad \text{(a)}$$

where the second equality uses the connection between $\chi_b(t-t')$ and $C_b(t-t')$, and the third arises from integration by parts and the fact (essentially a notational convention) that $f(0) = f = \sum_i c_i y_i = f_b(0)$. $f_b(t)$ differs from f(t) only as time evolves from the initial phase space point at t = 0.

Now consider the average of $f_b(t)$, averaged with the initial conditions x and $\dot{x}$ held fixed:

$$\langle f_b(t) \rangle \Big|_{\substack{x,\dot{x} \text{ fixed} \\ \text{at } t=0}} = \langle \ f_b(t; x, \dot{x}, \{y_i, \dot{y}_i\}) \ e^{\beta x f} \ \rangle_b \ \frac{1}{\langle e^{\beta x f} \rangle_b}$$

This equation follows from the fact that the distribution of initial bath variables differs by the factor (βxf) from that when the bath is uncoupled from the primary variable. Let us employ the y_i's that are the normal modes of the bath, where mode y_i has the frequency ω_i. Then we have

$$\langle f_b(t)\rangle\Big|_{x,\dot{x}\text{ fixed}} = \sum_i c_i \langle [y_i \cos\omega_i t + (\dot{y}_i/\omega_i)\sin\omega_i t]\, e^{\beta x c_i y_i}\rangle_b \frac{1}{\langle e^{\beta x c_i y_i}\rangle_b}$$

$$= \sum_i c_i \cos\omega_i t \, \frac{\partial}{\partial \beta x c_i} \ln \langle e^{\beta x c_i y_i}\rangle_b$$

$$= \beta x \sum_i c_i^2 \langle y_i^2\rangle \cos\omega_i t = \beta x C_b(t) \qquad \text{(b)}$$

where the second to last equality follows from the fact that each y_i is an independent harmonic oscillator variable and therefore obeys a Gaussian distribution. The last equality is true since for normal modes $\langle y_i y_j\rangle = \langle y_i^2\rangle \delta_{ij}$, hence

$$\sum_{i,j} c_i c_j \langle y_i y_j\rangle \cos(\omega_j t) = \sum_i c_i^2 \langle y_i^2\rangle \cos\omega_i t \; .$$

By combining Eqs. (a) and (b) with $m\ddot{x}(t) = f_0[x(t)] + f(t)$ and averaging over initial conditions of the bath variables, the generalized Langevin equation follows directly.

8.22 (a)

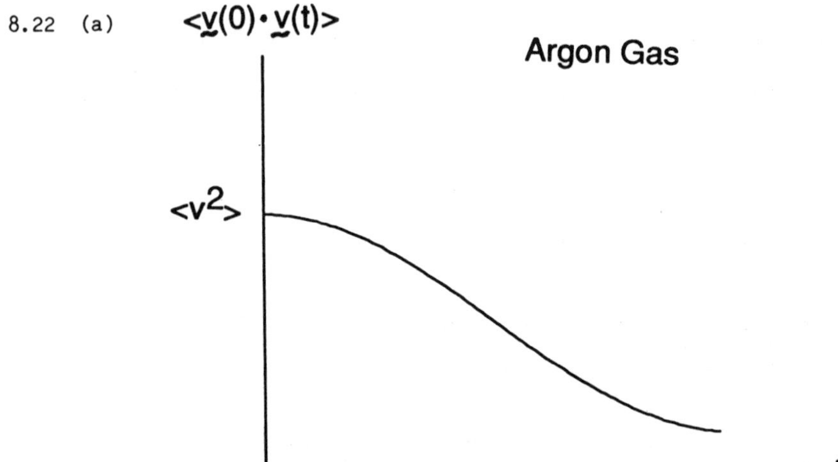

The very slow decay has a time scale of τ, the mean collision time in the gas.

(b)

$\langle v^2 \rangle$

T

t

Argon Solid

The oscillations persist for long times because for small displacements the atoms effectively sit in a harmonic well with period T . The slow decay is due to collisions with other atoms and also anharmonicities in the well which cause dephasing between different trajectories.

(c)

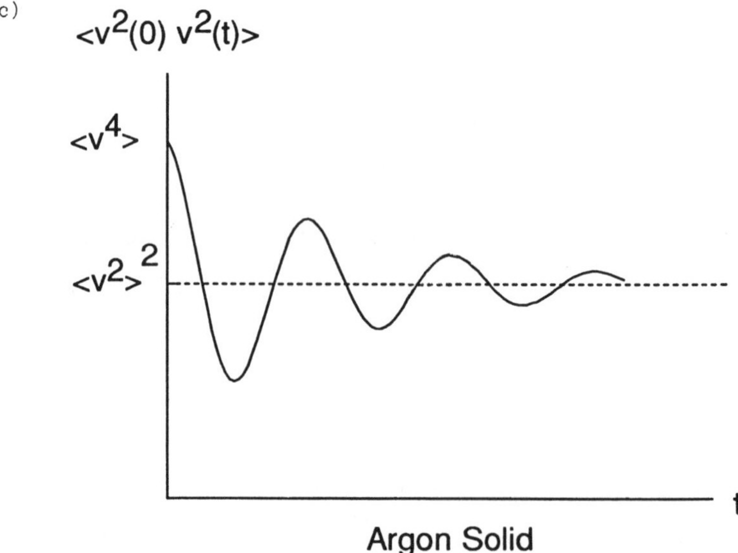

The decay in the oscillations is due to the processes described above. Note that even if the lattice were one-dimensional, the minima of $\langle v^2(0)v^2(t)\rangle$ would not drop to zero because the ensemble average is over various initial conditions, so the turning point times are averaged over. In three dimensions, we have orbits with

no turning points. This is reflected in the three separate phases that are averaged over:

$$\langle v^2(0)v^2(t)\rangle = \langle\left(v_x^2(0) + v_y^2(0) + v_z^2(0)\right)\left(v_x^2(t) + v_y^2(t) + v_z^2(t)\right)\rangle$$

$$= 3\langle v_x^2(0)\ v_x^2(t)\rangle \quad \text{in the case of an isotropic lattice.}$$

(d)

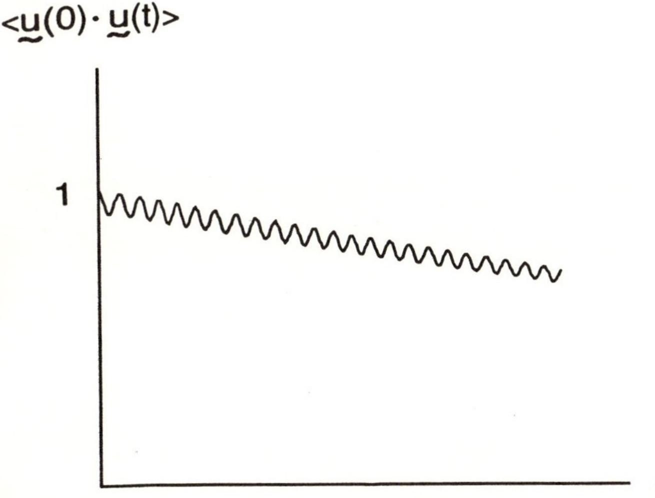

The small oscillations are due to the vibrations in the lattice. The long time decay is due to the infrequent process of a CO molecule flipping over completely in the lattice.

8.23 For a small perturbation from the equilibrium zero-field distribution, the fluctuation-dissipation theorem tells us the rate of relaxation to equilibrium is the same as that of a spontaneous fluctuation to that non-equilibrium distribution. At time t = 0, the field is shut off but the flow $\overline{v}(0) \neq 0$. Since the flow will dissipate with time,

$$\overline{v}(t) = \Delta\overline{v}(t) \propto \langle\delta \underset{\sim}{v}(0)\cdot\delta \underset{\sim}{v}(t)\rangle \propto \langle \underset{\sim}{v}(0)\cdot \underset{\sim}{v}(t)\rangle$$

or

$$\frac{\bar{v}(t)}{\bar{v}(0)} = \frac{\langle \underset{\sim}{v}(0)\cdot\underset{\sim}{v}(t)\rangle}{\langle v^2\rangle} .$$

The relaxation time is then

$$\tau_{relax} = \int_0^\infty dt \frac{\langle \underset{\sim}{v}(0)\cdot\underset{\sim}{v}(t)\rangle}{\langle v^2\rangle}$$

or $\tau_{relax} = \beta m D$.

8.26 Substituting $c_3(t) = c - c_1(t) - c_2(t)$ into the rate equations gives

$$\dot{c}_1(t) = -(k_{31}+k_{13})\, c_1(t) - k_{13}c_2(t) + k_{13}c$$

$$\dot{c}_2(t) = -k_{23}\, c_1(t) - (k_{32}+k_{23})\, c_2(t) + k_{23}c$$

or, in matrix notation,

$$\frac{d}{dt}\begin{pmatrix} c_1 \\ c_2 \end{pmatrix} = -\begin{pmatrix} k_{31}+k_{13} & k_{13} \\ k_{23} & k_{32}+k_{23} \end{pmatrix}\begin{pmatrix} c_1 \\ c_2 \end{pmatrix} + \begin{pmatrix} k_{13} \\ k_{23} \end{pmatrix} c$$

The homogeneous part can be solved by expressing $\begin{pmatrix} c_1 \\ c_2 \end{pmatrix}$ in terms of the eigenvectors of the above matrix, while the inhomogeneous part just adds a constant to d_1 and d_2: written in the basis that diagonalizes the above matrix,

$$\frac{d}{dt}\begin{pmatrix} d_1 \\ d_2 \end{pmatrix} = -\begin{pmatrix} \lambda_1 & 0 \\ 0 & \lambda_2 \end{pmatrix}\begin{pmatrix} d_1 \\ d_2 \end{pmatrix} \Rightarrow \begin{array}{l} d_1(t) = f_1 e^{-\lambda_1 t} + \text{const} \\ d_2(t) = f_2 e^{-\lambda_2 t} + \text{const}' \end{array}$$

where f_1 and f_2 are constants determined by the initial conditions. Thus the concentrations c_1 and c_2 can be written as the sum of two decaying exponentials,

$$c_1(t) = A_1 e^{-\lambda_1 t} + B_1 e^{-\lambda_2 t} + \langle c_1\rangle$$

$$c_2(t) = A_2 e^{-\lambda_1 t} + B_2 e^{-\lambda_2 t} + \langle c_2 \rangle$$

where $\begin{pmatrix} A_1 \\ A_2 \end{pmatrix}$ and $\begin{pmatrix} B_1 \\ B_2 \end{pmatrix}$ are eigenvectors of the above matrix.

(a) So,

$$\Delta c_1(t) = A_1 e^{-\lambda_1 t} + B_1 e^{-\lambda_2 t}$$

$$\Delta c_2(t) = A_2 e^{-\lambda_1 t} + B_2 e^{-\lambda_2 t}$$

where

$$\lambda_{1,2} = [(k_{31} + k_{13}) + (k_{32} + k_{23})]/2 \pm \sqrt{\{[(k_{31} + k_{13}) - (k_{32} + k_{23})]/2\}^2 + k_{13}k_{23}} \quad .$$

(b) For $e^{-\beta Q} \ll 1$, then k_{13}, $k_{23} \gg k_{31}$, k_{32}.

We can write

$$\lambda_{1,2} = [(k_{31} + k_{32}) + (k_{13} + k_{23})]/2 \pm \sqrt{\{[(k_{31} + k_{13}) + (k_{32} + k_{23})]/2\}^2 - (k_{31}k_{32} + k_{31}k_{23} + k_{13}k_{32})}$$

$$= [(k_{31} + k_{32})/2 + (k_{13} + k_{23})/2] \times [1 \pm \sqrt{1 - (k_{31}k_{32} + k_{31}k_{23} + k_{13}k_{32})/[(k_{31} + k_{13} + k_{32} + k_{23})/2]^2} \quad .$$

$\lambda_{1,2}$ is now conveniently written in terms of the small parameters k_{31}, k_{32}, so that we can expand the root $\sqrt{1-\varepsilon} \approx 1 - \frac{1}{2}\varepsilon$:

$$\lambda_{1,2} \approx [(k_{31} + k_{32})/2 + (k_{13} + k_{23})/2] \times \{1 \pm [1 - \tfrac{1}{2}(k_{31}k_{32} + k_{31}k_{23} + k_{13}k_{32})/((k_{31} + k_{13} + k_{32} + k_{33})/2)^2]\} \quad .$$

So the + root is

$$[(k_{13}+k_{23})/2 + O(\varepsilon)]\,[2 - O(\varepsilon)] \approx k_{13} + k_{23} = \tau_{\text{transient}}^{-1}$$

and the - root is

$$(k_{31}k_{32} + k_{31}k_{23} + k_{13}k_{32})/(k_{31}+k_{13}+k_{32}+k_{23})$$

$$\approx (k_{31}k_{23} + k_{13}k_{32})/(k_{13}+k_{23}) = \tau_{rxn}^{-1} .$$

From the above,

$$\tau_{rxn}^{-1} \approx k_{31}\left(\frac{k_{23}}{k_{13}+k_{23}}\right) + k_{32}\left(\frac{k_{13}}{k_{13}+k_{23}}\right) \ll \tau_{\text{transient}}^{-1} ,$$

and so the relaxation is dominated by τ_{rxn}. When $k_{23} \approx k_{13}$,

$$\tau_{rxn}^{-1} \approx (k_{31}+k_{32})/2 .$$

(c) As shown above, the faster transient decay occurs on a time scale of

$$(k_{13}+k_{23})^{-1} \sim k_{23}^{-1} \quad \text{if } k_{23} \geq k_{13} .$$

(d) The two decay rates are analogous to the two rates in the reactive flux description: $\tau_{mol} \approx \tau_{\text{transient}} \ll \tau_{rxn}$. The connection can be made by imagining preparing the system at the transition state, i.e., in state 3. Then the decay into states 1 and 2, $c_1(t)$ and $c_2(t)$, follow the two decay rates, one much faster than the other. But $c_1(t) \propto \langle \delta n_A(0)\delta n_A(t)\rangle$ by the regression hypothesis, and the time derivative of $\langle \delta n_A(0)\delta n_A(t)\rangle$ is just the flux in the reactive flux picture. In particular, τ_{rxn}^{-1} is on the order of k_{31} or k_{32}, and is the plateau value for the reaction rate.

(e) This is similar to the transition state theory idea where $e^{-\beta Q}$ is the probability of getting to the transition state 3, and $D \propto 1/\eta$ is the rate to cross the barrier (3) once there. As we showed previously

$$k^{(TST)} \propto \langle |v| \rangle \langle \delta(q-q^*) \rangle .$$

8.28 $T = 300$ K $\quad \eta = 0.01$ poise $\quad D_{Ar} = 1 \times 10^{-5} cm^2/sec$

$m_{Ar} = 40$ amu $\quad m_{H_2O} = 18$ amu

(a) and (b):

The velocity distribution is just the Maxwell-Boltzmann distribution and is independent of the potential. So

$$\langle v_{Ar}^2 \rangle = \frac{2}{m_{Ar}} \langle \frac{1}{2} m_{Ar} v_{Ar}^2 \rangle = \frac{2}{m_{Ar}} \frac{3}{2} k_B T$$

$$= 1.87 \times 10^9 \ cm^2/sec^2$$

or

$$\langle v_{Ar}^2 \rangle^{1/2} = 4.3 \times 10^4 \ cm/sec$$

for Ar atoms in the vapour and in solution.

(c) Assuming inertial motion in the vapour, the time to move 10Å is

$$t = 10Å/\langle v_{Ar}^2 \rangle^{1/2} = 2.3 \ psec .$$

(d) Since $\Delta R^2(t) = 6Dt$, the time to diffuse 10Å in solution is t = 170 psec. We are justified in assuming diffusive motion since 170 psec is long compared to the velocity relaxation time calculated in part (e).

(e) The relaxation of the non-equilibrium distribution will follow the same rate as that of spontaneous fluctuations, such as in $\langle \underset{\sim}{v}(0) \cdot \underset{\sim}{v}(t) \rangle$. The decay time for $\langle \underset{\sim}{v}(0) \cdot \underset{\sim}{v}(t) \rangle$ is given by

$$\tau_{relax} = \beta m D \approx 16 \text{ fsec} .$$

(f) From 8.24 $D = \frac{1}{2 \text{ or } 3} \frac{1}{\pi \beta \sigma} \frac{1}{\eta}$

So if η doubles, D halves, and the diffusion time $t = \frac{\Delta R^2(t)}{6D}$ doubles to 340 psec. In contrast, since the temperature is still 300K , $\langle v_{Ar}^2 \rangle^{1/2} = 4.3 \times 10^4$ cm/sec remains the same.